"*Move Your DNA* is one of the most important books I've ever read. If you've suspected 'nutritious' movement is as important to your health as nutritious food, you're right. This book explains why."
—Robb Wolf, author of *The Paleo Solution*

"One of my favorite books on movement."
—Perry Nickelston, DC, NKT, SFMA, author of *Stop Chasing Pain*

"Exceptionally well informed, informative, written, organized and presented, *Move Your DNA* is especially recommended to the attention of non-specialist general readers who are having to deal with mobility issues—especially as they grow older. It will also prove to be of immense value for both amateur and professional athletes, as well as concerned parents wanting their children to grow up whole and healthy."
—*Midwest Book Review*

"Biomechanics. Pelvic floor. Iliopsoas. These words may spark intrigue for fitness pros, but they're like kryptonite for much of the general population. That's where Katy Bowman comes in. Bowman succeeds with a teaching and training style that is equal parts education and entertainment."
—*IDEA Fitness Journal*

"I would recommend this book to everyone. Whether you've been doing the fitness thing for years or are an absolute beginner, you will find seeds of wisdom to help you move more and better. *Move Your DNA* is a paradigm shifter that acknowledges the complexity of modern movement habits while also providing helpful ways to improve and grow."
—*Breaking Muscle*

"This informative analysis of natural exercises uses apt analogies to demonstrate how best to build strength. *Move Your DNA* is enjoyable, convincing, and sure to change the way fitness buffs (and coach potatoes) move."
—*Foreword Reviews*

"Katy Bowman condenses outlandishly complicated cellular mechanics into words that you, your mother, and your children can comprehend. She helps you understand how your invisible components compose you…Once you read this important book, there is no turning back. You will never think about your human body, its habitat, its stressors, and your own humanity in the same way ever again."
—Jill Miller, Creator of Yoga Tune Up and author of *The Roll Model*

"Katy is an important influencer when it comes to promoting the transition from traditional regimented exercise to a more 'global movement' view, where the focus is on staying active throughout the day."
—Dr. Joseph Mercola

"As a Human Biologist, I find the ideas expressed in this book stimulating. The whole notion that we no longer move in the way our hunter-gatherer ancestors moved—the way that suits our evolutionary heritage—is quite intriguing and could justifiably form the basis for more concerted academic study. It could even form the basis of what might be called 'Body Use Theory.'"
—Stephen Lewis, Senior Lecturer in Biological Sciences, University of Chester, UK.

"Filled with scientific evidence, *Move Your DNA* is a compelling manifesto for natural movement. Katy Bowman does a tremendous job at explaining, in a clear and humorous manner, why we need to shift our perception from the reductive notion of 'exercise' to a refreshing, effective, movement-based lifestyle. The many practical drills in the book will help even the most 'out of shape' to get back to natural movements in gentle, progressive, and safe ways."
—Erwan LeCorre, creator of MovNat

"Katy teaches us a profound yet obvious fact about human health: Just as a boring assortment of six common foods cannot give us a truly nourishing diet, neither can a few mundane tasks and force-fed exercises keep us in shape. Our greatest vitality comes from living out the vagaries of Nature's whim—and if we can't do that, we can at least simulate the variety that we are adapted to."
—Samuel Thayer, author of *The Forager's Harvest: A Guide to Identifying, Harvesting, and Preparing Edible Wild Plants* and *Nature's Garden: A Guide to Identifying, Harvesting, and Preparing Edible Wild Plants*

"Wow! Katy clearly separates the 'ease' from the 'dis-ease' created by our captive lifestyles."
—Phillip Beach, DO, author of *Muscles and Meridians*

"*Move Your DNA*…is a pioneering and necessary book and an instant classic… [T]his book may provide a lightning bolt of insight for people who experience not just pain, but a range of physical symptoms or illnesses…In my experience as an osteopathic physician and holistic OB/GYN treating a full range of women's health issues including pelvic floor dysfunction and female pain, movement prescriptions are an essential part of any treatment plan or wellness program, and I am already recommending *Move Your DNA* to my patients and colleagues."
—Eden G. Fromberg, DO, FACOOG, DABIHM, Clinical Assistant Professor of Obstetrics & Gynecology at SUNY Downstate Medical Center

"Katy's depth of knowledge is simply astounding, as is her ability to communicate profound, life-altering information in a way that provides as many laughs as it does jaw-dropping, awe-striking facts. Katy has revealed the elemental importance of nutritious movement—the Whole Movement Movement!—and it is absolutely the missing piece to the puzzle of health and wellness. There are no words to describe the importance of the ideas within this book. They have revolutionized my practice and changed my life."
—Liz Wolfe, NTP, author of *Eat the Yolks*

ALSO BY KATY BOWMAN

MOVE YOUR DNA

KATY BOWMAN

FOREWORD BY JASON LEWIS

Printed in the United States of America

Second Edition, Eighth Printing, 2023
ISBN: 978-1-943370-10-8
Library of Congress Control Number: 2014939868

Propriometrics Press: propriometricspress.com
Cover and Interior Design: Zsofi Koller, liltcreative.co
Illustrations: Jillian Nicol
Cover images: Shutterstock, Alis3D
Indoor exercise and author bio photos by Jen Jurgensen, outdoor photos by Cecilia Ortiz.
Photos on pages 226 and 227 by Shelah M. Wilgus.

The information in this book should not be used for diagnosis or treatment, or as a substitute for professional medical care. Please consult with your health care provider prior to attempting any treatment on yourself or another individual.

Excerpt from Bernie Krause's, PhD, Jan. 31, 2001, speech before the World Affairs Council, San Francisco, "Loss of Natural Soundscape: Global Implications of Its Effect on Humans and Other Creatures" © 2014 Wild Sanctuary. Used with permission. All rights reserved.

Excerpt from *Nature's Garden: A Guide to Identifying, Harvesting, and Preparing Edible Wild Plants*, published by Forager's Harvest Press, is © 2010 Samuel Thayer. Used with permission. All rights reserved.

The Library of Congress has cataloged the original edition as follows:

Bowman, Katy.
    Move your DNA : restore your health through natural movement / Katy Bowman ; foreword by Jason Lewis.
       pages : illustrations ; cm
    Subtitle from cover.
    Issued also as an ebook.
    Includes bibliographical references and index.
    ISBN: 978-0-9896539-4-7

    1. Exercise therapy. 2. Human locomotion--Physiological aspects. 3. Sedentary behavior--Health aspects. 4. Health. I. Lewis, Jason, 1967- II. Title.

RM725 .B69 2014

615.82   2014939868

For my children for demanding firsts, my husband for accepting seconds,
and the grace of God that there were thirds.

ACKNOWLEDGMENTS

A body is made of cells, and a book is made of ideas. My cells were gifted to me by my parents to fashion as I saw fit, and in this same way, ideas and talents have been gifted to me throughout my life—the work here is the culmination of the thoughts of many.

I thank my mother, especially, for drilling it into my head that I was the captain of my own ship even when my embodiment of this idea did not make things easier on her as a parent. I thank my father for bequeathing me kindness, which is helpful in packaging radical information for the masses. To my siblings, especially my youngest sister, KLD, I acknowledge your ability to still love me after years of my constant and loud chatter. Growing up with a KatySays is no easy feat, and I love you for doing so.

My children are my muses and the literal offspring of passions upon passions, but my husband is the true alchemist in this story. You've given my life context, my words meaning.

And speaking of words, this book would not have been "this shape" without the passions and talents of superstar editor Penelope Jackson. There's no way this book could have been if you hadn't studied the content for a couple of years beforehand. Which speaks not only to your caliber, but also to how you're being ridiculously underpaid at this point.

Great thanks to Zsofi Koller and Agi Koller for the skills you've generously shared with me, not only on this project, but on all our visual media of late, to Stephanie Domet for your assistance and organization, and to Debbie Beane for your work every day.

I am very much obliged to Dr. Jeannette Loram for lending her expertise in cellular biology, most specifically her gracious coaching of the most appropriate use of language surrounding DNA and genetic expression.

The images created for this book are not just pictures but an extension of the text. For these storytelling graphics I owe one life debt to illustrator Jillian Nicol and photographers Cecilia Ortiz and Jen Jurgensen. Also a huge thanks to Tim, Michael K., Breena,

Michael C., John, Theresa, Angeliese, Crystal, Delia, Galina, and Gayle for modeling the original edition's correctives.

Theresa, this book could not have been done without you. In fact, I don't know how I managed to live so long without you in my life. Thank you for being there and for always listening, closely, to the spoken word. You're a superlistener, a skill trumped only by your thoughtfulness.

To Jason Lewis and Tammie Stevens of Billyfish Books, thank you so much for gracing this book with your incredible story. To Dr. Seth Horowitz, Dr. Bernie Krause, Dr. Steve Lewis, and Sam Thayer, thank you for sharing your words and expertise so willingly. I hope thousands search out the iceberg of knowledge living below the tips you've allowed me to present here.

All the major players working on this book live their lives on their own terms. And because they embody their message—a message threaded throughout this book—the energy behind *Move Your DNA* is hugely passionate. For anyone out there living their dream, I thank you for doing so. Your contribution to this fabric we are all a part of is just as important to my life as this book might be to yours.

METTA, KATY B.

A NOTE ON THIS EDITION

One of the things that intrigues me the most about movement is the science of it. This is why I wrote *Move Your DNA* in the first place: I wanted to establish more thorough definitions and variables when thinking and talking about movement. I originally thought of *Move Your DNA* as an *exercise science* book rather than an *exercise* book. Still, with over seventy exercises and alignment adjustments peppered throughout, the first edition of *Move Your DNA* was also a book of exercises.

Because one big idea in *Move Your DNA* is "exercise less and move more," the exercises in the first edition were presented not as a routine, but as moves to put into your daily life here and there. Yet many readers wrote to me asking for more thorough exercise programming—now that they were motivated to move more, they wanted routines to start using ASAP. So I decided to offer an expanded edition of the book augmenting the exercise component. Thus this edition includes new photographs redone for more clear instruction on form, a bullet-pointed exercise glossary, and three exercise routines you can use during your "exercise time."

I also took the opportunity a new edition affords to here and there add new and pertinent findings in mechanotransduction and exercise science research. I hope the whole book beckons us outside to move more, if only a little bit at a time, through nature.

TABLE OF CONTENTS

FOREWORD

In 2013 I was recognized by Guinness World Records as the first person in history to circumnavigate the planet by human power. A rather fancy way of saying I walked, cycled, and inline skated across the Earth's landmasses, and kayaked, rowed, and swam across its oceans, rivers, and seas. It was a 46,505-mile journey that took thirteen years of my life. I was not only defined by this act but also shaped by it, even more than I had previously realized, as I have since discovered through Katy's unique insight.

I learned one thing very quickly after pedaling away from the Greenwich meridian line all those years ago: No two environments on the face of this Earth are identical. Every swell, every road, every mountain, and every minute spent crossing the "same" stretch of desert created a new experience, for both my mind and my body.

While the physical obstacles often seemed insurmountable—a crocodile attack, two broken legs, malaria, altitude sickness, septicemia, and more—it was the mental stamina that ultimately carried me across the finish line. The freeing of my body, physically, from society—the norms, constructs, and expectations—required an equal releasing of my mind.

I had no experience as a so-called adventurer—far from it; I ran a window-cleaning business. Before I began my journey, I did no training, had no experience at sea, had never held a kayak paddle or worn a pair of skates. I simply decided to take a leap of faith and allowed my body to adapt to my mind's decision. Our bodies are capable of amazing feats if our minds agree to cooperate.

Because my expedition was considered an epic physical accomplishment, my readers often feel they can't relate to me as much as they would like. They think I must have exceptional mental and physical stamina, but I assure you, nothing could be further from the truth. I'm no different from you.

While pedaling my little boat *Moksha* across the Atlantic, I experienced a profound moment of what I can only describe as samadhi. This experience impacted me greatly, as it left me no doubt whatsoever that we are all one. We are not separate from nature or the animals; nor are we separate from each other. Our interconnectedness is undeniable. Over the years, I've met people from all walks of life, each of them experiencing their own physical and mental challenges. Whether we are Buddhist or Baptist, African or American, work as a ditch digger or dentist, despite fatigue, injury, loss of hope, and sometimes sheer desperation, we are all doing our best to adapt to whatever life throws our way, then attempting to wake up and do it all over again.

For me, leaving the "comforts" of modern-day society and adjusting to a much more rudimentary life generally centered around survival was actually easier than the more recent years spent writing my book. Now I am often parked at my computer screen for many hours at a time. I'm no longer living as our ancestors did, directly under the sun and moon, experiencing that life-enhancing, primal connection with nature, but existing under man-made lighting and the unrelenting stress we have all sadly become accustomed to, caused by a constant barrage of technology, meetings, meals on the go, pollution, and traffic, with the accompanying cacophony of mental and literal noise that goes with it.

Krishnamurti said, "It is no measure of health to be well adjusted to a profoundly sick society." I for one intend to follow the teachings of another wise sage, and invite you on a new journey with me as we all begin to change our minds and move our DNA.

–JASON LEWIS

INTRODUCTION

The pattern of disease or injury that affects any group of people is never a matter of chance. It is invariably the expression of stresses and strains to which they were exposed, a response to everything in their environment and behavior.

CALVIN WELLS, *BONES, BODIES AND DISEASE*

W ho wants to be healthy? I know I do. But the term *healthy* means different things to different people. When it comes to true, objective health, how do we know if we have it or not?

Health has traditionally boiled down to either how you look (healthy as a horse!) or how you look on paper (passed my lipid panel with flying colors!). We tend to give much less consideration to *how we feel*. But how you feel is the earliest indicator of your health on a cellular level.

You might feel pretty good most of the time. But perhaps you find yourself unable to move once or twice a year because you throw your back out. Maybe you get headaches with enough regularity that you keep aspirin in your desk. Or deal with chronic constipation. Do you have a "trick knee" or a habit of spraining your ankles? Do those knees keep you from taking long walks? How are your natural, biological functions—like digestion, elimination, and sleeping—working for you? Is your life peppered with little inconvenient health glitches?

To get a more objective picture of your health, take out a piece of paper and write down the following:

- Clinical diagnoses received over your lifetime
- Every prescription medication you take and why
- Every over-the-counter medication you take, and how often
- Every surgery you have had or need to have
- Visits to the hospital, doctor, chiropractor, or any other allied health professional
- Body parts that "alert" you regularly or on a semi-regular basis
- Body parts that hurt
- Body parts that aren't working to the best of their ability
- Health issues you worry about having in the future

You probably have items to list for almost all of those bullet points, and most of your friends and family do, too. So why is health—as measured by how we feel, by what our bodies are telling us, by *how our bodies are functioning*—eluding us? What's going wrong?

We've made huge advancements in the areas of antibiotics, sewage treatment, and vaccines, but developed or wealthy areas around the world share a family of health issues nonetheless. Not the communicable diseases found in areas lacking progressive medical science—diseases that used to be a human's worst enemy—but rather ailments linked to lifestyle. Often referred to as the *affluent ailments* or *diseases of affluence*, this list includes, among many others: coronary heart disease, metabolic disorders (like Type 2 diabetes), certain cancers, osteoarthritis, osteoporosis, allergies, depression, obesity, hypertension, asthma, and gout.

But the term *affluent ailment* is misleading, as it implies that these issues arise from an excessive amount of money and the resulting lifestyle more money brings. More recent data shows an emergence of these "affluent" ailments in poor countries and communities, where extra money certainly isn't the issue. The more likely culprit, it turns out, isn't necessarily an abundance of wealth or the extra time that comes with extra money, but the physical environment created by globalization, urban dwelling, new social structures, and technology.

I'd like, therefore, to adjust this term, as the word *affluent* is both inaccurate and unhelpful. Categorizing diseases by the passive condition of living in a "good" place, instead of by the way you *act* in a place or time in which you live, implies the location is the cause of the diseases. In most cases, modern environments do not *prevent* us from adopting behaviors with better health outcomes. We *choose* to drive instead of walk. To push our kids in a stroller and not carry them in our arms. To push our food in a cart and not carry it on our back. We slouch into our furniture, and let our shoes support our feet. Yes, our modern culture of convenience appeals to our human instinct to conserve energy, but we are not imprisoned in any real physical way. And because we are not being forced into the office, the designer footwear, or the supercomfy couch, I suggest here that we replace *affluent ailments* with *diseases of behavior*.

You don't need a lot of money to make a disease of behavior. When the basics of life (food, clean water, and shelter) are so easily obtained, nature takes over. It is completely natural to avoid work (movement, in this case) when physical engagement is not required—in other words, when the consequence of being sedentary is not immediate death. Diseases of behavior arise in situations where the quality of the food being consumed is poor, stress levels are elevated frequently, and the work performed by the body is low and unvarying (as in a non-exerciser) or high and unvarying (as in those doing repetitive tasks, manual labor, or pursuing what we call "fitness" in the typical manner).

Despite our great fortune to live in a time when we aren't at great risk for communicable diseases, we are, in fact, dying—slowly, in bits—from our natural tendency to do as little as possible. Our unquenchable desire to be comfortable has debilitated us. Ironic, as there is nothing comfortable about being debilitated. This paradox—that advancements to make our lives less physically taxing have taxed us physically—is profound and has led to an emergent scientific hypothesis: Perhaps the only way out of our poor physical state, created by our culture of convenience, is a return to the behaviors of our ancestors.

MOVEMENT, OUTSOURCED

Before we lived in the age of convenience, movement of the human body was necessary for sustaining life. Finding, capturing, and collecting food and water required all-day, lifelong frequencies of movement. Seeking and creating shelters required strength and stamina. Propagating the species required a healthy, moving body for ease of copulation, gestation, and delivery. At one point on the human timeline, movement and all its variables we associate with physical health—endurance, strength, and mobility—were necessary for survival.

Over the last ten thousand years, most humans transitioned from a migratory, hunter-gathering population to living in sedentary farming communities, then industrialized nations, and then our current technology-based culture. You and I dwell in a time when movement has been almost entirely outsourced. A moment on our phone can secure food, delivered right to the door. We can seek shelter on Craigslist from the comfort of our chairs. Heck, we can even find a mate online these days, securing a partner without flexing anything but our fingers on a keyboard. While the abundance of food and money varies around the globe, for almost all populations, the current global environment has changed in at least one way across the board: Moving is not required.

EXERCISE LESS, MOVE MORE, MOVE BETTER

Move Your DNA presents a new paradigm of movement. Because DNA can be expressed differently, depending on how external factors impinge upon the cells within which the DNA resides, and because movement is one of these factors, the way we move directly influences how our bodies are shaped—for good and ill. It is not enough for me to tell you just to "move more." You also need to "move better" if you are to enjoy a more sustainable state of well-being.

This is a serious call to movement—serious, but not unpleasant. Thousands of readers and students of mine have found the physical, psychological, and emotional shift that comes with this material to be profound and delightful.

Most people have very little idea about how movement works in our bodies, or how much movement is required for natural biological function. It is not my intention to make you freak out about your health, though I'm aware that I might. My highlighting the essentialness of movement should be used to create opportunity for healing (a positive response) rather than fear of illness (a negative response). Many people are shocked when they realize just how easy it is to move more (note I said *move,* and not *exercise*) and how radically better they feel by making tiny skeletal adjustments throughout the day. Are you ready? Let's do this!

THINK

NUTRITIOUS MOVEMENT
AND DISEASES OF CAPTIVITY
CHAPTER 1

We see in order to move; we move in order to see. WILLIAM GIBSON

Once, when I was in college, I went for an entire day without eating. I wasn't planning to fast, but I had a hundred-page paper due the following Monday and sat down to type the entire thing on a Friday. I worked straight through for twenty hours before I realized, when I dropped, exhausted, into bed the next day, that I hadn't eaten a bite or had a sip to drink. My lack of eating that day wasn't a big deal, but the next morning my body was sending me some serious YOU NEED TO EAT signals.

I'm sure most of you have had a similar experience of missing your regularly scheduled food intake due to travel, work, kids, school, or something else that simply got in the way. Maybe you've even chosen to fast for a period of time. Regardless of the reason, the physical signal following a period of food-abstinence is usually hunger. Which makes sense, right? Eating is a physiological requirement. Food—specifically, the nutrients found in food—is not optional.

Nevertheless, eating optimally can be a challenge. Say, for example, I describe the best, most nutritious diet in the world, and dictate that it must include:

adequate calories (energy), an appropriate ratio of macronutrients (fat, protein, carbohydrates), an appropriate quantity of micronutrients (vitamins, minerals, organic acids, trace minerals), and enough fiber. Furthermore, it must be fresh and free of harmful chemicals.

Luckily, most of you reading this book aren't slowly starving or unable to purchase food, so we can all move toward a healthy diet using this template for nutrition. Yes, our pocketbooks might be tight enough to prevent purchasing all optimal ingredients, but with a little prioritization, we can usually figure out how to score the food we want by letting the stuff we don't need go.

But I'll bet that many of you reading this book have probably done quite a bit to educate yourself on a nutrition profile even more detailed than I've listed above. Take "an appropriate ratio" of fats, for example: What kinds of fats are necessary? Saturated? Monounsaturated? Trans fats? WHAT ABOUT OMEGA 3s? WHY AREN'T OMEGA 3s ON THE LIST?

I'll bet you understand that, when it comes to diet, details and context matter, and that my template, while a good start, isn't very thorough. Take "adequate calories" for example. If you need to eat 2,500 calories a day for adequate energy, is eating 2,500 calories' worth of Snickers bars an adequate diet? Of course not, right? What if, every day, you ate 2,500 calories' worth of non-GMO, fresh-from-the-farmers'-market oranges? Are you healthy now? What if you ate 2,500 calories' worth of beef liver every day? Healthy yet? A basic and valid food guideline like "total calories" can be erroneously applied without more detailed criteria.

FOOD (AND MOVEMENT) ESSENTIALS

Keeping in mind the incredible number of details necessary for building a nutritious diet, let us consider another type of input: movement.

I propose that movement, like food, is not optional; that you have been receiving signals of movement hunger in response to a movement diet that is very low in terms of quantity and poor in terms of quality—meaning you aren't getting the full spectrum of movement nutrition necessary for human function. Chances are, you are either lacking movement nutrition entirely, or you are eating mounds of movement Snickers without ever reaching for a movement kale salad.

Movement and food nutrition are both incredibly nuanced, far more than we give them credit for. In grade school, most of us learned about specific diseases arising from a single missing nutrient—vitamin C was eventually identified as the culprit behind sailors' scurvy—but other than that, very few of us can list every macro- and micronutrient, their specific functions, and how each relates to all the others and to our health. I was reading *Dancing Skeletons*, a book by nutritional anthropologist Katherine Dettwyler about her time working in Africa, when I found a section about kwashiorkor, a severe form of malnutrition common in young children throughout the tropics. The hallmark diet of this disease is high in calories (from sweet potatoes or other starches) but low in protein. In this case, the low protein is not the problem—other children who eat equally low amounts of protein but fewer *total* calories are not likely to develop the disease. It's the ratio of the nutrients that contributes to the development of kwashiorkor.

This section of Dettwyler's book resonated with me because I recognize that the outcomes of an exercise program depend largely on the ratio of all the movements to each other. Exercise (a repetitive intake of an isolated muscle contraction to fill a hole of missing strength) is often prescribed like vitamins (a capsule ingested to decrease a nutritional void). One of the arguments I am most known for professionally is that the way the Kegel exercise is prescribed can actually be harmful and not helpful at all. A Kegel is like a starch in the case of kwashiorkor: when done excessively and in the absence of other movement vitamins, it can create a negative outcome—too much pelvic-floor tension. The Kegel (as I'll expand upon in chapter 10) is not inherently more "bad" than a sweet potato, but neither is a sweet potato (or Kegel) health-making when consumed in isolation.

. . . .

KEGEL EXERCISE

A contraction of the pelvic floor often prescribed to prevent the leakage of urine when coughing or running.

. . . .

"Good" nourishment, whether you're talking about food or movement, cannot be reduced to a single or even a few variables, and bad diets or exercise programs are not the result of the deficit of a single component. When you're eating a well-balanced diet (in every sense of the word) there is a *sum total* effect of wellness that permeates the entire body. Every individual nutrient serves a unique role in the process, and many nutrients create site-specific effects in the body. Often the ailments you have—of the nails or hair, the liver, the eyes—can be tracked to the specific nutrient you are missing, so you start to think of foods that contain that nutrient as medicine. But a different point of view is that we aren't really sick—we are just starved. And we aren't using food as medicine. Food isn't medicine at all. It's just food; we need its nutrients to survive and thrive. It's that simple.

MOVEMENT NUTRIENTS

Both food and movement create a cascade of biochemical processes that alter the state of your physiology. The conversion of movement "input" to biochemical processes is called *mechanotransduction*.

With apologies to those with robust training in biology, allow me to give you a brief introduction to the organization of the human body. The way academics have organized the body on paper for easier study is this: Your body is made of organ systems which, in turn, are made of organs. These organs are made of tissues which, in turn, are made of cells.

MECHANOTRANSDUCTION

The process by which cells sense and then translate mechanical signals (compression, tension, fluid shear) created by their physical environment into biochemical signals, allowing cells to adjust their structure and function accordingly.

But really, your body is just made up of cells, each of these cells being connected to each other via a network of extracellular matrix—a complex network of polysaccharides and proteins that provides structure as well as regulates all aspects of cell behavior. When you move what you probably think of as your body—arms, legs, torso, head—you are not only rearranging the larger structures of your limbs and vertebrae but also your small, cellular structures.

We experience load 100 percent of the time. Gravity is one force your body responds to constantly. Just as your body would collapse if it didn't have bones, cellular organs within your cells would fall in response to the gravitational force if the cytoskeleton weren't there to hold them in place. But even though the gravitational force is constant here on Earth, the loads created by gravity depend on our physical position relative to the gravitational force. For example, gravity is always working on your bones, but the load created by gravity differs depending on how those bones line up with the perpendicular force of gravity. A month of the horizontal positioning common to bedrest can decrease your muscle and bone mass. Same gravitational force. Same genes. Different position. Different body.

And gravity isn't the only force that loads our cells. Simply put, a force is a push or pull on an object. In the case of our bodies, many of the objects being pushed and pulled are our cellular sensing organs, which is how we feel the universe around us. External pressures (like the interaction between bone, muscle, and a chair), frictions (like a new pair of shoes against your foot skin), and tractioning forces (remember, in eighties movies, those old-school pulley-on-cast devices used in hospitals after someone broke their leg skiing?) all create cellular deformations within our body, as does movement itself. The lengthening and shortening of larger tissues like muscle creates pushes and pulls on the small-scale stuff.

Most of us understand that our bodies respond to mechanical input. Our optometrist monitors us for high pressure in the eyes to avoid damage to the optic nerve. We're familiar with pressure wounds like bedsores developing in those who sit or lie continuously without shifting much. We discuss, with ease, the new pair of shoes that made blisters at first, and how that one time we had a cast, our muscle wasted away, leaving a noticeable difference. We are quite comfortable with these examples (I hope), but most don't ponder how these phenomena occur. Why, exactly, does the optic nerve die in a high-pressure environment, leaving us with glaucoma? Mechanotransduction is, finally, being researched as the underlying mechanism of many diseases. Diseases of mechanotransduction are those ailments arising from an area of cells (then tissue, then

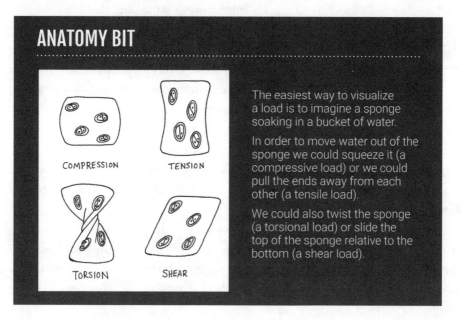

ANATOMY BIT

COMPRESSION

TENSION

TORSION

SHEAR

The easiest way to visualize a load is to imagine a sponge soaking in a bucket of water.

In order to move water out of the sponge we could squeeze it (a compressive load) or we could pull the ends away from each other (a tensile load).

We could also twist the sponge (a torsional load) or slide the top of the sponge relative to the bottom (a shear load).

ON LANGUAGE AND TELEOLOGY

Teleology means attributing purpose to physiological mechanism. Because the skeleton, for example, adapts very much to what we choose to do, to imply conscious or purposeful design is to ignore the somatic process. It is easier, though, when writing about the body, to write, "The hip is designed to carry the weight of the torso," than to write, "A consequence of the hip and pelvis orientation is that the hip has become robust enough to carry the weight of the torso." If ever I do use the term *designed to,* it is for the purpose of keeping the writing smooth and easy to follow. Whether a particular alteration is evolutionary or somatic (self-induced), the notion of purpose is a concept that is moot.

organ) troubled by the mechanical environment you've created, both directly and indirectly.

Movement, position, and the resting state of our musculoskeletal system are huge influencers of our mechanical environments. While we think of movement as something we do to train our bodies into better shape, most don't consider how that "better shape" comes about. Well, now you know. It is via the process of mechanotransduction that our physical self adapts (in shape) to our experience of the physical world.

More precisely, the physical expression that is your body is the sum total of loads experienced by your cells. Picture, for a moment, standing in a huge forest. A wind blows through this forest and you look up to see trees bending this way and that way. Some trees hardly move at all and some rustle quite a bit. The quantity and qualities of tree movement depend on the direction of the wind, its strength, and how long it lasted.

It's important to remember that the load *is not the wind.* The load is the *effects* created by the wind. The load is how the trees physically experience the wind. Every tree experiences the wind uniquely, depending on its height and girth, its position relative to other trees (maybe the effects of a wind are made small due to taller, surrounding trees), and many other factors. Also, the tree does not experience the wind similarly over the entire tree. Areas of the trees with branches might catch more of the wind, bending the trunk more in those areas. In other places, without branches, the wind might have little effect beyond a slight pressure on the bark.

You are used to thinking of yourself as one big body, and not the sum total of many tiny parts. When we think of loads—especially loads in a book about exercise—we tend to apply the term to the force ("I was loaded down with a heavy weight") instead of thinking about how that heavy weight created unique deformations (and loads) on a trillion of your parts. The twenty-pound weight is not the load. The load is the experience created by carrying it.

NOW, LET'S TALK ABOUT LOADS

Imagine you are standing in the center of a trampoline. The weight placed on the trampoline is equal to your weight, but due to the structure of the fibers that make up the trampoline, the load created by your weight is not experienced equally by all parts of the trampoline. There are areas that are deformed more than others. The area just below your feet bulges toward the ground the most, with the amount of distortion decreasing as you look outward toward the frame of the trampoline. There are connective materials—springs—that connect the trampoline to the frame. These springs also experience the load, as does the frame itself—although the distortion here is much less. Invisible, almost.

When you stand on the trampoline you load the entire structure, but every part of the structure experiences the load differently. I had you imagine how a trampoline would be deformed with you standing in the center. Now, imagine standing in a different spot. Although your weight is exactly the same in both cases, the trampoline's experience of your weight is entirely unique, depending on where you stand. And you're only standing in this scenario. I could have you jump. Or run across the trampoline. Or backflip! In each of these cases, the trampoline's experience of the loads created by your movements is entirely unique, nanosecond to nanosecond.

It's easy to comprehend the importance of load on the final shape of a trampoline. Imagine now that your tissues are trampolines inside of you. The loads your tissue-trampolines experience are the result of your positioning—both when you're still and when you're moving.

LOADS ARE LIKE SNOWFLAKES

Every unique joint configuration, and the way that joint configuration is positioned relative to gravity, and every motion created, and the way that motion was initiated, creates a unique load that in turn creates a very specific pattern of strain in the body. Every load experienced by the body, whether the distortion is created by our activity (or lack thereof), the position of that activity, the impact of that activity, or the repetitiveness of that activity (or lack thereof), is its own "nutrient"—what I will now refer to as a *load profile*.

Here is a thirteen-pound pumpkin.

These are two pumpkins that, together, weigh thirteen pounds.

This is a group of pumpkins that, together, weigh thirteen pounds.

On a scale, each of these pictured pumpkin groups creates a thirteen-pound weight. But these "equal" weights do not create the same load within the body. The way your body adapts to the loads placed upon it has less to do with the weight (thirteen pounds, in this case) and more to do with how you carry it.

Here are five different ways one can carry the thirteen-pound pumpkin.

Here are more unique ways to carry thirteen pounds of weight.

And finally, how to hold thirteen (inconvenient) pumpkin pounds.

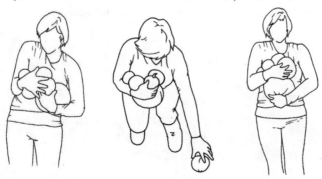

These pictures of the many different ways you can carry thirteen pounds of weight are an example of load variance. In one case, the pumpkin could require the flexion of my wrists and elbows, causing the muscles of my arms and back to contract. In another, my arms are free from working but the vertebral disks in my neck might be deforming. In some cases I work one side of my body more than the other; in others the work is distributed more evenly.

The takeaway here is that every load is a unique cellular deformation (the movement equivalent of a nutrient), even when the applied force (thirteen pounds of pumpkin) is exactly the same.

WEIGHT VS. LOAD

"It's not the load that breaks you down, it's the way you carry it." I adore the sentiment of this Lena Horne quote, even though I'd love to modify it to say, "It's not the *weight* that breaks you down, it's the load created by the way you carry it." But with the modifications it's not very poetic (and probably confusing), so I'll leave it as is. You get what it's saying, right?

Whether you're talking pumpkin or body pounds, "weight" does not describe what kind of load is created. Often people are told that their injuries have been caused primarily by their body weight. Perhaps you have a bad knee and have been told that your weight has created a higher-than-normal load. Perhaps your doctor explained that your extra weight has pressed on your knees in a way that has rubbed all your cartilage down to the bone. The solution? Reduce your weight, thus changing the load (true), and heal your knee. But how are you supposed to drop thirty pounds if your knee doesn't articulate without pain?

Loads are often oversimplified to "weight" because it makes them easier to understand, but there is much more going on with your sore knee (or foot, or

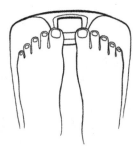

back, or pelvic floor) than your weight. There are millions of people without any extra weight experiencing the same disease or injury and many with extra weight not experiencing it as well. Weight is not the be-all and end-all of loads. When you want to improve your health, it's much more important to consider *how you carry* your weight than to spend hours contemplating the lone data point that is Your Weight.

For example, the way your foot is positioned while you're walking can create higher-than-normal loads within the knee. The more your feet turn out when you walk, the higher the loads placed on medial knee structures like the anterior cruciate ligament (ACL) and the medial meniscus. Do you need to lose thirty pounds to improve your knee loading? No. You can just start reducing the turnout of your foot.

THERE ARE LOADS OF LOAD VARIABLES

Once you understand load profiles, you can begin to better evaluate the effect of exercise on health. Does exercise create the loads we need to stay healthy? Are all exercises equally beneficial to all of the body? What makes up the most nutritious movement plan?

Let's look at biking research. Cycling athletes tend to have lower bone density than running athletes. Why? Because sitting on a bike creates less of a vertical load than carrying your weight on your legs does. A cyclist certainly creates loads by pushing and pulling on the pedals, but the loads from running or walking are different from the loads experienced during cycling. If you want your bones to stay dense enough to carry the load of your vertical weight, you must actually load them with your vertical weight. Cycling does not create bones strong enough to hold a vertical load, although they would be strong enough to handle the loads created during cycling.

Of course, all cycling does not have the same effect on a body. Off-roading cyclists, with body parts subject to bumpy, jolting trails, and who stand more on their bike, fare better, bone-wise, than their smooth-road riding pals. Even those little jolts to the body make a difference to your cells.

In the field of exercise, "bike riding" can be a specific category of fitness-making. But we've seen here how road biking vs. mountain biking can result in different health outcomes (like bone density). You can break down every type of exercise to the loads your body experiences during that exercise to find out if the perks of a favorite fitness activity come with any adaptations that detract from long-term health goals.

When it comes to fitness, we are typically most concerned with, "Did you ride the bike or not?" But when you evaluate movement or exercise in terms of cellular outcomes, you have to be more specific. *How* did you ride it? Fast? Slow? Uphill or down? And what was your upper body doing? Leaning on the handle bars? What about your pelvic goodies? Were they pressing on the seat? Or does your seat have a cutout (in which case the higher pressure is now surrounding your junk and not directly on it). No matter the activity, when it comes to health, of utmost importance is the loads created.

Every rate, size, and angle at which a force is being applied creates a unique environment for your cells. Just as we do with nutrition, we can keep breaking down a load profile into smaller and smaller bits. In tissue-injury research, variables of force application that round out a load profile include magnitude, location, direction, duration, frequency, rate, and variability.

Here is a list of the different ways the process of loading affects the outcome. The first three are those variables affected by the weight and the geometry of the body. The other variables have to do with the timing and spacing of the loads you create in your life.

Magnitude: The amount of force applied. (Imagine the sponge full of water. Did you squeeze the crap out of it, or did you squeeze it just a little bit?)

Location: Where on the structure the force was applied. (Did you pinch a little corner of the sponge, or wrap your hand around the entire thing?)

Direction: How the force was directed. (Did you squeeze it? Or pull the ends away from each other? Twist it? Push the top towards the bottom? Squeeze the sides together?)

Duration: The time interval over which the force was applied. (For how long did you squeeze the sponge?)

Frequency: How often the force was applied. (Did you squeeze the sponge six times or fifty-seven times today?)

Rate: How quickly the force was applied. (Did you take thirty seconds to slowly wring out the sponge, or did you try to wring it out in a single second?)

Variability: Whether the magnitude of the force stayed constant or changed over the interval of application. (If you squeezed the sponge for thirty seconds, were you squeezing it hard the entire time, or were there a few seconds where you eased up a bit?)

EVEN MORE ON LOADS

The timing and rates of loads are important because loads are *occurrences over a period of time*. Each load has a beginning, middle, and end, and can look different depending on when you observe it.

If you put a balloon on a chair and sit on it, the force applied to the balloon is eventually equal to your weight. But the deformation of the balloon is gradual, because your weight doesn't magically appear on the balloon—your muscles lower your mass into the chair over a period of time, maybe a second. The lateral bulging of the balloon is most noticeable once all of your weight is on the balloon, but the balloon experiences a smaller deformation at first, and then a little more and a little more and a little more, until finally it is fully "squished" by your weight. If I evaluate the magnitude of the force placed on the balloon just before it pops, it would be closer to your full weight, but if I evaluate the load on the balloon at any point before that, the force applied is less.

And human tissue comes in more than one type. Each type of tissue reacts differently to different variables; an identical set of forces will create a different load depending on what tissue type you're pushing on. Just as pressing hard on a stone would result in a different outcome than pressing just as hard on a balloon, you get different cellular deformations in different tissues.

Even though the tissues respond differently, they are all connected, which means that a load you perceive as only happening in one part of your body is actually affecting all other parts of you, and affecting each part uniquely. I like to say that you're not just putting a backpack onto your body. You're actually putting on a trillion backpacks—one on each cell—but the way each cell carries the pack depends on how far it lives from the actual backpack. That cell down there in your heel bone? Yeah, it feels the backpack too, although not as much as the cells in your shoulders.

But it's not just about "feeling the weight." The angles of deformation created by a force aren't always what you think they will be. Imagine tugging on the bottom of your shirt. Or better yet, go ahead and tug on your shirt right now. Pull the bottom hem toward the floor. You can probably see vertical wrinkles, slightly angled wrinkles, and areas with no wrinkles at all. You might see, if you pay close attention, that the sides of your shirt move toward the middle. Applications of a single downward force create deformations (read: loads) in entirely different directions than you might intuitively think.

ACTIVE MOVEMENT AND OUR LACK THEREOF

Armed with your new load expertise, let's talk about how cellular loads are an inherent part of movement.

Many people view their musculoskeletal system as a bunch of levers and pulleys that move the body around the planet, but there's so much more happening.

1. Your skeletal muscles fire, moving you about the planet.
2. The action of muscles firing compresses embedded arterioles (the blood vessels just off the arteries), causing them to open, which pulls blood into working areas.
3. The movement of blood into now working areas pulls oxygen into the area, providing fuel for the cells, and at the same time pushes out cellular waste that is constantly in production.
4. The loads created through movement can both *displace* and *deform* your body. Firing your biceps muscle group to rotate your lower arm at the elbow to lift a pumpkin is an example of displacement. Displacement occurs when a part (in this case the arm bone) does not change shape, but moves as a whole. Only, the arm bone holding the pumpkin isn't a really solid or rigid lever; bone is soft. Unless you're a robot (apologies to the Six Million Dollar Man and Bionic Woman), your tissues are soft. All of them. Which means that holding the pumpkin also creates small *deformations*—these are a change in the original shape of a tissue. These tissue shape changes are mostly microscopic—the traction of an object in the hands, the pull of the hand skin on the connective tissue just below, and the minuscule bend of an arm bone under the weight of the pumpkin are all examples of "invisible" loads that don't come with the huge eye-candy provided by displacement, but nevertheless create a cellular experience.

In order to better understand how your body's shape is brought about by mechanotransduction, we need to start thinking small. The activity we see with our eyes (the changing position of the elbows and knees, feet, and hips) is not only an indicator of whole-body movement, it is also an indicator of more subtle movements in the system, tissues, and cells underneath the skin.

At this point it should be easy for you to see how our lack of movement input is slowly suffocating us on a cellular level. Motions that used to be incidental to living (read: occurring all day long) and cellular loads that used to be built into everyday life have been doled out—to computers, machines, and other people moving on our behalf. There is no way to physically recover the specific bends and torques, no way to recreate one hundred weekly hours of cell-squashing in seven, and no technology, at this time, smart enough to override nature. Illness is typically looked at as physiology gone wrong. I assert here that in most cases, our physiology is responding *exactly as it should* to the types of movement we have been inputting. Instead of thinking of ourselves as broken, we should recognize our lack of health as a sign of a broken (mechanical) environment.

"But wait," you say. "I do exercise, so what about me?" The prevailing understanding of exercise includes the belief that exercise *of any type* improves the distribution of oxygen to all tissues, but this is not the case. Movement of any type improves the circulation (read: oxygen and waste removal) *only through the muscles that are being used for that specific movement*. Even if you're a great exerciser—maybe you bike or jog religiously—only the muscles you've used for that specific exercise garner any benefits. Over time, heavy use of your body in one particular pattern makes strong tissues next to weaker ones, which creates an environment where an injury can slowly develop.

The frequent consumption of varied movement is what drives essential physiological processes. Movement is not as optional as we have led ourselves to believe. Just as a lack of food (or, heaven forbid, oxygen) leads to a multitude of biological signals and physiological outcomes, people are living in their body-houses surrounded by screaming alarms in the form of pain, illness, and disease, and they are unaware of the source of the problem. You have been doing the movement equivalent of under-eating and under-breathing, which is having an impact on your whole body, right down to the cellular level.

Of course, diet, stress, and environmental factors can all change the expression (or the physical outcome) of your DNA. But it is my professional opinion as a biomechanist that movement is what most humans are missing more than any other factor, and the bulk of the scientific community has dropped the ball. With respect to disease, the human's internal mechanical environment has been the least-discussed environment of all—a staggering oversight when almost every cell in your body has specialized equipment *just to sense the mechanical environment*. Kinesiologists and kinesiology departments—a field of academia supposed to, by definition, be studying human movement—have replaced movement with fitness and sports (more on this in chapter 3). Within the scientific field of anatomy we've created a hierarchy of structure that implies the cardiovascular and nervous systems are not only separate from but much more impactful on our health than the musculoskeletal system.

Every single thing our bodies do requires movement—initiated by our musculoskeletal system—to be performed with ease. Digestion, immunity, reproduction—all of these functions require us to move. You can eat the perfect diet, sleep eight hours a night, and use only baking soda and vinegar to clean your house, but without the loads created by natural movement, all of these worthy efforts are thwarted on a cellular level, and your optimal wellness level remains elusive.

As I'll explain later, natural movement is not just about muscles and exercise. (And just to be confusing, sometimes natural movement isn't even about moving.) There are many essential *passive loads* that come from interacting with all the Earth's forces. But allow me to use a different species to explain essential (and passive) loads, as sometimes we are too close to our own cultural experience to see beyond it.

DISEASES OF CAPTIVITY

Have you ever been to an aquarium or sea-life theme park featuring an orca (also known as a "killer whale")? Or maybe you've watched *Free Willy*. Either way, you might have noticed the collapsed dorsal fin of this breed when in captivity.

Marine biologists and sea mammal veterinarians have dug deeper to find out why this Flaccid Fin Syndrome (FFS, although marine biologist Wende Alexandra Evans points out that the fin is in fact rigid, and "Folded Fin Syndrome" would be more accurate) occurs, asking what it is about captivity, specifically, that increases the chance of this happening. They have created a list of behaviors unique to captivity:

1. Orcas in captivity swim only in a counter-clockwise circle.
2. Orcas kept in shallow tanks miss out on moving through the higher static fluid pressure environment created by oceanic depths.
3. The diet of an orca in captivity is different—the food has lower water content—than in nature.
4. The amount of time an orca spends at the surface is greater when in captivity.

Marine scientists observed that fin deformities are also found in natural settings, only at a much lesser frequency and to a smaller degree. You might see a little fin curl, but not to the degree or in the direction that is found in most captive males. The ocean-dwelling whales that did have floppage also seemed to have some sort of trauma associated with the dorsal fin (as indicated by visible wounds), and some whales with the deformity demonstrated an improvement when spotted at later dates. There are no records of a captive orca's FFS improving.

There were other data points to consider:

1. A natural softening (increased wobbliness) occurs in the fin during growth spurts in pubescent orcas.
2. The males have longer dorsal fins than females.
3. The dorsal fin of an orca is made of a tissue similar to our ligaments. Mostly collagen fibers, the dorsal fin does not have any bones or contain muscle to hold itself, which makes it a passive tissue.

Putting all these data points together, scientists hypothesize that the most likely cause of FFS is the mechanical environment of captivity—in this case the *missing loads* to the fin that would have been created by a lot of forward swimming at depths that pushed the passive tissues of the fin into an upright position, and also the *unnaturally high loads* created by single-direction, tight-circle swimming, and the fin's increased exposure to gravity (due to the fin being above the surface of the water). Whales that would be most susceptible to FFS would, then, be those whales with the tallest fins, whales that did their growing in captivity, and whales with a genetic precursor for wobbly collagen.

So now, my point.

It seems like a dorsal fin, a structure necessary for stable swimming—an essential life task for a whale—should come with some sort of stabilizing mechanism (like that provided by a muscle) to prevent problems like fin floppage. But if we consider this problem from an evolutionary perspective, why would a whale, evolved to eons of swimming in a certain way, through an ocean environment, need a stabilizing mechanism if the simple act of swimming in that environment creates the necessary forces for function? A stabilizing mechanism would be unnecessary hardware for the whale, costing energy to use and maintain. The whale's equipment is perfectly suited for swimming; it is only the captive whale's environment that is the problem. Even for the whales with a genetically "weaker" fin, the genetics are only poor *in the specific environment of captivity.*

Human diseases are repeatedly explained to us in terms of their chemical or genetic makeup; meanwhile, we've completely ignored the load profile that the function of our body depends upon. As far-fetched as this may sound, we, like these floppy-finned orcas, are animals in captivity, and our tissues are not suited to the loads created through the way we move in our modern habitat.

GENES AND DISEASES OF BEHAVIOR

Let's say that, in the future, all whales exist only in captivity, and captive whales are the only ones we can observe, gather data about, and research. Over time, the high frequency of flopped-over fins will appear to be the norm. Without other, wild, whales to compare them to, we would probably focus on a particular chemical or genetic makeup that would make FFS more likely to occur. We'd assume that male whales, whales with long dorsal fins, and whales with a certain gene for "type X" collagen were all at risk of FFS simply because of those factors.

If all whales were swimming in tanks, then the *way* they swam would not be an obvious factor in the development of FFS. This is how all whales swim, right? All the whales we've measured swim exactly this way. I mean, it's not like the whales aren't getting exercise. Doesn't all swimming use the same muscles? Don't we make sure that the whales get between one and three hours of swimming a day? We've got that covered, so there must be some *other* reason.

But, of course, all swimming is *not* equal. Even when the muscle groups used are the same, the forces placed on the whale's body throughout its life can vary greatly with variations in speed, habitual position relative to gravity, etc. The physical outcome the whale experiences depends on loads. When we view movement as a series of loads, it is easier to understand why swimming slowly in circles at the surface of the water for a couple of hours a day is not the same as straight-ahead swimming, at great depths, punctuated with sprints—the natural swimming-foraging and mating behaviors found in wild populations.

Trade floppy fins for bum knees, collapsed arches, eroded hips, tight hamstrings, leaky pelvic floors, collapsed ankles, you name it—and consider our load profile. Walking on a treadmill an hour a day creates an entirely different load profile than walking over the ground for an hour. Wearing shoes to walk that hour creates a different load profile than walking without. Spending the eight hours before and after that hour-long walk sitting down creates a different outcome than dividing an hour's walk throughout the day.

The loads we experience in today's world differ hugely from the loads people experienced a hundred, a thousand, and ten thousand years ago. Yet we blithely accept that our health issues—which so many of us share—are genetic. *Genetic*, a term we've internally defined as *beyond our control*. Whether out of convenience or ignorance, we have failed to address the habitat in which our genes dwell, and the impact of the way we move on the state of our health.

MOVEMENT, LOADS, AND YOUR DNA
CHAPTER 2

Nature isn't classical, dammit, and if you want to make a simulation of nature, you'd better make it quantum mechanical, and by golly it's a wonderful problem, because it doesn't look so easy. RICHARD FEYNMAN

I've spent most of my academic career, almost twenty years now, studying human biomechanics. I am endlessly fascinated by animal cells, the human body, and movement, but sometimes, after I've spent ten hours thinking about body biomechanics, I need a break. On days like these, I'll make a huge cup of tea, draw a warm bath, and settle down for a relaxing read of…plant-mechanics research.

Allow me to share with you my favorite tree-mechanic factoid: Trees are shaped by the wind. Seriously. I mean, tree genes specify the primary shape and color and texture of a tree. These genetically determined outcomes help a biologist classify a tree as a redwood or a manzanita, etc., but it is the movement of a tree, specifically the all-day, every-day stimulation created by wind, that dictates the girth of a tree's trunk and branches as well as how often and at what angles a tree branches. Doesn't that just blow you away? (What I find most interesting about this is, how does a tree store this mechanical input? Trees grow so high and for so long, they have to store the mechanical information in a way that allows future growth to capitalize on mechanical input. WHERE IS THE TREE'S BRAIN? If you know, dear reader, please do not hesitate in emailing me this information. I will be eternally grateful.)

GENES AND VIEW-MASTERS

After that plant-mechanics-and-genes refreshment break, let's talk people and genes. If you attended high school in the last hundred years, you were probably presented with a cellular model that basically states a cell's nucleus contains all the information necessary for cellular replication, with the genetic information (DNA) determining a cell's behavior. Thusly the state of every tissue (made up of cells) and every organ (made up of tissues) and every system (made up of these organs) is dictated by our genes.

But as the trees have just shown us, genes are not that simple. Scientists have observed that a person simply having a particular gene doesn't automatically create a particular outcome. This means, for example, that you and your neighbor could both have a breast cancer gene, but only one (or neither) of you gets cancer. The fact that identical genes can behave differently depending on environment has led to an emerging field of study called *epigenetics*, a branch of biology studying how a cell's environment can affect the behavior of the cell itself.

A *gene* is a specific sequence of DNA on a single chromosome that encodes a particular product. Many people associate genes with the concept of *pre-determination*, and will use these terms interchangeably, as in, "The doctor said my bad knees were genetic," or, "Research shows that cardiovascular disease is genetic." But using the term *genetic* in this way is at best outdated and at worst totally paralyzing to the person with the issue.

It would be more accurate to think about genes as "range-setters" of an outcome. Your genetic constitution is not a picture of how you are going to look now and in the future. Rather, your genetic makeup is like one of those disks you put in a View-Master—a plethora of potential outcomes for you to select by toggling on the View-Master's lever.

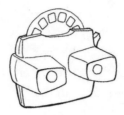

THE MECHANOME

You are probably familiar with the term *genome* (your genes, gene modifiers, and the "junk" in between), as it is used regularly when it comes to health and human experience, but chances are you have never heard the term *mechanome*.

The mechanome, in the example of the View-Master, would include all the forces and machinery necessary to move the lever and advance the disk. The machinery, the process of creating stimulation, the loads that are perceived by your cells' mechanosensors, and the response triggered by the cell deformations—are collectively called the *mechanome*. A mechanome is the interplay between forces and biology.

For example, your genes contain information about the ratio of muscle fiber types you have, which affects *the potential* for your muscles to develop in response to exercise—for instance, whether you'll ever be able to be a world-class sprinter—but genes do not run the programs for developing your body into an athlete's. Rather, this development occurs when you create stimulation through your actions. If you (and your genes) lay in bed for fifteen years beginning the day you were born, you would not end up looking the same (in person or on paper) as you would have had you (and your genes) been upright and moving around for fifteen years. This is an extreme example, but all movement and lack of movement create subtle differences in outcome in individuals and their genes.

BIASED TOWARDS THE GENETIC PERSPECTIVE

Science education can take a long time to catch up with science. Even today, decades after scientists achieved a better understanding of the role environment plays in genetic expression, new anatomy and physiology students are still presented with the "nucleus controls the cell" model; the more advanced cellular model is only taught in graduate-level biology classes specifically for mechanists—mechanobiology.

The result of this oversight is that the doctors we depend on are typically not trained in the science of mechanotransduction. (And really, there is no place to add mechanotransduction to medical curricula. Doctors already have to go to school forever to be able to recognize an emergency, as well as the rarest of all pathologies, in all humans. It's ridiculous to expect the medical community to also be responsible for learning math and physics in addition to biology, chemistry, anatomy, physiology, and how to deal with humans in pain. It's time to give them a break and take some personal responsibility. There, I said it.)

MECHANOBIOLOGY

A relatively new field of science that focuses on the way physical forces and changes in cell or tissue mechanics contribute to development, physiology, and disease.

Regardless of why, the mechanical environment is perhaps the most important, and yet certainly the most ignored, aspect of a cell's environment. Mechanotransduction, as you know by now, is the process by which cells sense and respond to mechanical signals. You also know that, through loads, mechanical signals are being created 100 percent of the time, both by our movements and by how we are positioned when we're not moving. Movement (not only exercise, but every gesture, big or small, made by the human body) loads the body's tissues, and the body's cells. Every cell contains a rigid network called a cytoskeleton, similar in function to our bones. Most recent findings in cellular biomechanics show that the deformation of the cell itself, and the load placed on the cytoskeleton, affect each cell's behavior, including how the cell regenerates.

Rigorous study of mechanotransduction phenomena is "new"—most of the research has occurred in the last two decades—but it was common knowledge in scientific circles over a hundred years ago, thanks in part to German anatomist Julius Wolff (of Wolff's Law fame).

Today, there is a large volume of scientific research regarding the effects physical loads have on the ailments and injuries we develop. Still, the greatest allocation of resources (and magazine headlines) focuses on genetic pre-determinism and biochemical markers (like high cholesterol in the case of heart disease). Despite scientific understanding that virtually all cells adapt to accommodate their mechanical environment and that biochemical signals for genetic expression might not even be necessary (it appears the cytoskeleton can directly transmit mechanical signals to the DNA via recently identified "cytofilaments"; see the Jorgens reference), our physical experience is repeatedly presented as an event that has little to do with our choices—such as how we have used our bodies since birth.

Our general lack of awareness of the mechanome should not muddle the fact that many of the processes occurring in the body, including genetic expression, can be regulated *mechanically*. When you understand this, you quickly see how searching for a health solution without considering your "movement environment" inevitably produces results that are limited in scope and benefit.

Recently, thanks to breakthroughs in nanotechnology, scientists can better see how cells transfer forces to each other, and how loads create adaptations in individual cells that result in a tissue outcome. Understanding the process should eventually help clinicians recognize many common health issues (including such diseases as osteoarthritis, osteoporosis, cancers, and collagenopathies) as *diseases of mechanotransduction*—and, more importantly, aid scientists in designing more specific load-intervention therapies. Until then, we can consider how we load our bodies every minute of every day and make changes, as necessary, right away.

WOLFF'S LAW

Wolff's Law was originally a set of mathematical equations used by anatomist Julius Wolff to predict the specific trajectory of bone formations. While the details (his mathematical model likening bone to an inflexible tissue, for example) proved inaccurate, the principle—that bone adapts to the loads created through mechanical usage—is still a fundamental tenet of osteology today.

Wolff's Law is a generalized term for the idea that changes in bony formation, resorption, balance, turnover, and remodeling space depend on body use—both as a forming juvenile and, less widely understood, as an adult. Anthropologists today use Wolff's Law as an underlying assumption that differences in bone morphology can be used to investigate differences in past mechanical environments.

YOU ARE HOW YOU MOVE

I don't know you personally, but chances are we share a similar developmental experience, which means our cells have had very similar mechanical environments. You were probably born in the hospital, then driven home in a bassinet in the back seat of a car. The first six months found you mostly lying down, and when you were taken outside, it was typically on your back, in some sort of stroller or pram.

The fine motor skills in your hands were cultivated most; you were given rattles and other objects to hold and manipulate. To develop leg strength, you would have required someone to hold you upright regularly so your legs could bear a little of your body weight—but this practice was deemed unsafe due to the scientifically unsupported but common belief that it causes babies to become bowlegged. And while your parents were told they shouldn't facilitate your standing, no one stopped them from constant swaddling, putting you in Happy Jumpers, Joyful Walkers, or other gleefully named devices that lead to scientifically demonstrated detrimental outcomes (like poor motor develop-ment or developmental hip dysplasia). Your natural reflexes of rolling over, sitting up, crawling, and then walking probably occurred, respectively, at around six, seven, eight, and twelve months.

YIKES

According to Active Healthy Kids Canada's 2013 Report Card on Physical Activity for Children and Youth, Canadian teens between the ages of fifteen and seven-teen walk an average of eleven minutes a day.

Once you were toddling, or maybe before, you were given shoes to "support" your feet, and you explored your world—until it was time to sit down, in your high chair, on your tricycle, or on your very own child-sized chair.

Your walking turned to child-running—first an awkward, stiff-armed gait, and eventually something more like "real" running. Hard to tell with those bulky diapers pressing your legs apart. Because you were heavy to carry or hard to control, you might have spent a fair amount of time in the stroller even when you weren't sleeping.

Starting school around age five, you found yourself squirming in a chair for hours on end. Sitting still was not in your nature, but after a couple of years, sitting still in your chair would be your most-practiced skill, trumping time spent reading, writing, playing games, and physical education in school. Like a ninja of sitting, you practiced sitting still in a chair more than any other activity, with hours and hours and hours in training, with no other learned activity even coming close in time spent practicing.

I'm not sure about you, but I played a lot after school, sitting on my bike or climbing on old farm tractors and staying outside until the sun went down. I did not have regular homework, certainly not the daily regimen today's kids have. As I grew older, my playtime slowly faded and was replaced with hanging out with friends. It was tons of fun, but it never involved much movement.

Maybe you sat and played an instrument, practiced a sport after school, were diligent in dance class, or wrote for the school paper. Chances are, your "after school" movement dwindled to an hour or two or became excessively structured—repeating the same drills over and over, in a sort of high-frequency/low-variety program.

Now that you're an adult, the chair and/or computer probably rule your life before, during, and after work. The bulk of your caloric intake is collected from the grocery store, ready to eat. If you are a good representation of "average," you drive almost everywhere and take at least one medication on a regular basis as well as a few painkillers a month, and have sought professional service for at least one musculoskeletal issue—probably your lower back. Your feet have been in shoes almost every waking hour of almost every day of your entire life.

If you are part of the some 40 percent of the United States population that exercises on a regular basis, you likely take your movement indoors, three to four days a week, for about forty-five minutes. You use some sort of device, machinery, or repetitious pattern, and you might be accompanied by loud music. There's a strong likelihood that your exercise involves moving your legs a lot but not moving your body much relative to the ground. Walking, a coordinated muscular event, is naturally reconciled with the continuous stream of visual input (a.k.a. optic flow). You move forward, but at the same time objects move past you. All of this data—the amount your joints move and muscles contract, and the rate objects are displaced in your field of view, are all integrated by your sensory system. Now that you're fixed in place on a treadmill, your brain is forced to adapt to large motions that take you nowhere, as indicated by data gathered by your eyes.

This movement timeline is a generalization, but there is a very strong chance that you, the reader, identify with almost all of it. Your movement timeline is crucial, as your body has been literally shaped by your movement experience. And when I say *movement*, I mean more than just exercise. I'm talking about every motion your body has made and every position it has maintained throughout your lifetime.

Imagine your body is made of clay, with each type and frequency of movement shaping the physical outcome. Take your imaginary body-ball-of-clay through your personal movement timeline, considering your early development, favorite activities, accidents or sports injuries, footwear habits, the desks at school, your favorite couch, and driving posture. Create the "resultant" shape in your mind. Now go look in the mirror. The molded clay in your mind should look like you, right now in the mirror. What you have done to date has resulted in your "shape." And remember, thanks to our understanding of loads and epigenetics, we know that the *literal* shape you are in affects not only the function of your body's tissue, but also your cellular health. In short, it affects everything.

HUNTER-GATHERERS, PAST AND PRESENT

The term *hunter-gatherer* refers to a member of a nomadic group of people who survive in the wild by obtaining food through both hunting and foraging. The term is a broad one, as it includes historical populations that subsisted exclusively via these means, as well as modern populations that hunt and gather some of the time while farming or storing foodstuffs other times.

Some of the time, I am referring to the historical hunter-gatherer populations and the data derived from looking at artifacts and assuming the conditions of the times. In other cases, I am using the term to refer to modern peoples who currently live this way and data gathered through direct interaction. It's important to note that modern hunter-gatherers are not relics of the past but modern peoples, affected by globalization.

GROWING UP NATURALLY

Let us consider another clay body, this one shaped by an all-nature, all-the-time childhood. Of course, nobody knows for certain what ancestral hunter-gatherers did all day, but we can hypothesize based on existing evidence. We can use physical and anthropological data to estimate how much a tribe could migrate over an average year, and imagine how they would have acquired food. Although modern-day hunter-gathering populations are not by any means living fossils, it can be helpful to see what their days and lives entail, especially when we're trying to integrate some of the data we already have. Even considering all the conveniences ancient hunter-gathering populations did not have—and they had almost none—can help us gain an idea of what their bodies had to do all day long.

As you can imagine, physical development for people nomadically moving through an all-natural setting is very different from that of people in modern societies.

PALEO MOVEMENT AND MISMATCH THEORY

Natural, or "paleo," movement is not the Paleo Movement. Over the last decade, a scientific hypothesis called the evolutionary mismatch theory has been gaining momentum, positing that the conditions under which human beings developed are similar to those necessary for biological function; that human physiology has adapted to certain stressors, the absence of which results in disease; and that our current environment (or lifestyle) is a mismatch for our physiology.

The word *paleo* has become a blanket term for the idea that, through cultivating ancestral practices, we can optimize our health, not because behaviors of yesteryear are quaint, but because they contain necessary inputs for humans. Natural movement is a branch within the Paleo Movement; other branches include diet, community, parenting style, etc.

These ideas are trendy but not new. The Cynics were an ancient group that rejected social values and argued that natural actions were a necessity for humans. Followers of Cynicism took their meals without cups or "manners," took care of bodily functions in public (the original term Cynic implies "like a dog") and had a devotion to "natural life." Socrates's disciple Antisthenes is documented as the founder of Cynicism, but Diogenes is said to exemplify it and to have died from eating raw octopus while attempting to prove a point about the unnaturalness of cooking. Point taken, Diogenes.

If you were part of an ancient hunter-gatherer tribe, your development would have resembled this scenario: Following an entirely unmedicated birth, you, a hunter-gatherer baby, were breastfed, slept with your parents, and exercised multiple times daily. You reached standing and walking milestones at the time many modern kids begin crawling. Exclusively held, you began "core exercising" with every step your parents took (all outside), your body position shifting minute to minute as you or your holder required, allowing you to explore both the world and an infinite number of loads via varying positions.

Just before age two, you were play-gathering, with repeated squatting and standing, digging and clambering, for hours a day. When not play-gathering, you played in constantly varying terrain. This all-day movement (and variability to movement) developed the skills, strength, and *shape* you would eventually need in order to function as an adult, and your gait and walking patterns were much less toddler-like and wobbly because you didn't wear diapers. Your pelvis and hips took the shape necessary to continue squatting, sitting on the floor, and walking a ton, and were not influenced by kid carriers, car seats, or continuous time in a single position.

Shortly after puberty, probably age fourteen, you were a fully functioning member of a tribe, participating in the same all-day movement patterns as your parents and walking medium (three-mile) to long (ten-mile) distances most days of your life. You walked every day and worked hard harvesting and carrying enough bounty to ensure survival. The frequent, weight-bearing loads of walking maximized your peak bone mass during the most crucial period of young adulthood.

As an adult, you don't exercise regularly, or ever, really. Instead, you use your body to get life done. Your total movements, varying joint positions, and rate of energy expenditure for a day's survival work easily exceed those found in a standard athletic workout. And in addition to moving more, you also relax frequently. You don't have the stress of driving, constant noise, constant information, and excessive light.

Imagine reconfiguring your clay body into this shape. A shape brought about by nature.

WHAT DOES YOUR BODY SHAPE SAY ABOUT YOU?

Anthropologists and medical researchers alike have concluded that the way humans move now is drastically different from how humans have moved over the bulk of the human timeline, which is easy to see once you compare your own physical timeline with that of someone in a traditional hunter-gatherer society. The physical requirements of the human body—the loads that drive many of the functions we depend on for living—are not well met by the quantity *and types* of loads created in a modern society.

Most cells depend heavily on mechanical stimulation. The loads placed on the body via movement translate into loads on the cells themselves, which creates cellular data, and it is at this level that change—in the form of strengths, densities, and shape—occurs. We use the word *disease* to imply that something has gone awry in our bodies; but as I said before, more often than not, our bodies are simply responding normally to the input they're given. Movement provides information for the body. Movement is an environmental or epigenetic factor. Our movement environment has been polluted, so to speak, and we've got the bodies to match.

YOU'RE NOT OUT OF SHAPE

Your body is never "out of shape"; it is always *in* a shape created by how you have moved up to this very moment. It is constantly responding and shifting to a continuous stream of input provided by your external and internal environments, even if that input consists only of sitting still, for hours on end.

A quick read of the hunter-gathering life above will give you a sense of how your daily movement measures up. Just imagine your own life and then begin stripping away the obvious things, like your living-room and dining-room

EVERYONE WANTS TO BE A HUNTER

I believe that humans have unique food and movement requirements. While we have similar foundational necessities, our intrinsic uniqueness calls for input that fuels and replenishes the way we use our physiology to complete our community job. As with any group of animals, every participant fills a different role—a role that capitalizes on individuals' strengths. Without a blend of strengths, our species would be vulnerable to gaps in our functionality. Said another way, not everyone needs to be the warrior. Not everyone a nurturer. Not everyone a hunter. Not everyone a gatherer.

My twenty years in the health and fitness field have shown me clearly that some people love (require, even) the demand of intense physical training. There are also many I've worked with who wished they loved to exercise, but don't. Despite how our love for physical work differs, we all have in common the need for fundamental movements and loads—the loads that are not dependent on our constitution or roles, and are similar across the board.

There are movement nutrients we are all missing due to our similar cultural experience. If you're an athlete, you won't find the most athletic feats of the hunter-gatherers listed here; I'm assuming you're already getting those loads, but might be experiencing some deterioration because you're performing these feats outside of a natural (meaning "as found in nature") context. You might be all run and no walk. Or you're doing two hundred squats a day, but in shoes, and without squatting facets in your bones (see chapter 10).

My point is, everyone—even the hunteriest of the hunters—was a gatherer first. When hunter-gatherers are children, their job is to gather. All are successful in this way first. To go forward, we must go back to figure out which movement basics we have failed to practice and which tissue adaptations we still need.

The beauty of these basics is that they establish a community-wide playing field where we can all grow together.

furniture. Quickly figure out how much time your
body spends in a position that uses the furniture's
"energy" for support. The body of the hunter-gatherer spent most of the time supporting itself.

Imagine how many times a day you would squat
for elimination purposes without your toilet.
Then walk into your kitchen and see how much
time you spend standing in front of the sink or preparing food at counter-level
instead of crouching or sitting on the ground. The more you can imagine just how
your activities of daily living differ from those found in nomadic, food-searching

BONE ROBUSTICITY: MECHANOTRANSDUCTION IN ACTION

While each of our bones has a genetic shape that is consistent enough to be recognized—Hey! That's a thigh bone!—the nuances of a bone are based on how that bone
is used over a lifetime (just like a maple tree, shaped by the wind!). Physical anthropologists have used bone robusticity to calculate the loads and movement patterns
used by earlier humans. Skeletons of people who rode horseback many hours, for
example, have a particular shape and density compared to those of people who didn't.
And when you move your body less, or rearrange your body parts (think of habitually
stooping forward) in a way that decreases the vertical load to bone, the pelvis and
thigh bones respond by becoming less strong. There isn't a problem with the bone-
building process in these cases; you've simply decreased the load to which your body
responds. That your body decreases tissue in response to decreased load is an indication of its metabolic savviness. Why would you expend energy maintaining tissue
you don't use? "Use it or lose it" is physiologically sound advice.

Whether you move or not, your choice stimulates the cytoskeletons of all your cells in
a way that signals "this behavior is what I do, so please adapt." The current epidemic
of osteoporosis—specifically loss in the wrists, ribs, spine, and head of the thigh
bone—speaks volumes about how we move. Our cultural patterns of localized bone
loss (osteoporosis is not typically a body-wide failure to make bone, which should be
a *huge* red flag for bone-disease researchers) are what you would expect from the
loading patterns we have in common. Skeletons are a sort of "living story" you are
continuously writing. Bone robusticity is a result not only of genetic data, but also of
data created via behavior. It is through your choices of movement and the cellular
loads these choices create that your body becomes your autobiography.

individuals, the easier it will be for you to understand why it takes more than an hour of exercise a day, however high in intensity, to recreate the distinct loading profiles of this lifestyle.

I encourage you to spend an entire day taking note of everything that you do, like turning on the faucet for water and opening the fridge for food. Imagine what you would have to do differently if these things didn't exist. The more you tally your behavior, the more you will realize just how your movement quantities compare to quantities humans are capable of moving. You'll also have a great list of where to start making small changes to your active loading habits.

PASSIVE LOADS MATTER TOO

The way we've come to erroneously equate fitness with health has interfered with our understanding of the body's dependency on specific loads. Maybe you got your workout in, but what about the loads you created the rest of the day? For how many hours a week is a chair pressing on your hamstrings? How does this constant pressure affect the blood vessels running down to your feet or the nerves in the pelvis? How do you sleep? And I don't mean how many hours; I'm referring to the loads created by the position in which you sleep. Have your body's tissues atrophied to the point where they are no longer able to adapt to a different mattress or pillow? This is a sign that the smaller joints in your body have stiffened to the point where doing nothing—on a pile of fluffiness—is too hard on your body (à la Princess and the Pea). Had you been sleeping in nature, this bedtime adaptation would not have settled in your cells, making you too weak to go without your pillow. I know *sleeping without a pillow* doesn't sound like CrossFit, but a great deal of our population would find themselves stiff and sore the day after a pillowless night simply because they've used their body in a new-to-them way.

There's another way in which modern loads are different that is almost never considered critically: underwear. Bras and underwear remove the gravitational load to our swinging and hanging bits. Breasts and testicles depend on gravity to load their suspensory muscles and ligaments, which in turn keeps these suspensory systems strong and able to move as necessary.

In men, the cremaster muscle lifts and lowers the testicles depending on their temperature. When they get cold, the cremaster pulls them into the body for warmth; when they are too warm, the cremaster relaxes them away to help dissipate heat. If clothing holds testicles close to the body all of the time, not only is the strength of the cremaster diminished, the temperature of the testicles is unnecessarily high all of the time.

In women, breasts depend on their suspensory systems and the muscles beneath them for support—systems that now adapt to the loads created by the bra, not to the weight of the breast. If your breasts are heavy, there is a greater chance you've been wearing a bra longer, which means that not only are your breasts heavier than others', your suspensory system is also weaker, relatively speaking. Interesting, as there may be a link between larger breasts and breast cancer. While researchers are looking to the genome for these answers, I'd like to wave a flag for the mechanome and suggest that they consider the Parable of the Fin of the Orca and load mismatch (see the sidebar on page 35 for more on mismatch).

Bra-wearing large-breasted women experience loads created by gravity almost identically to bra-wearing small-breasted women. If a bra renders the motion of both large and small breasts negligible, large-breasted women are at a greater mismatch between the loads their breasts require and the loads they experience—in the same way an orca with a tall fin is at a greater loss when exposed to unnatural loads.

With recent increases in both breast and testicular cancers—cancers that have roots in mechanotransduction—it is extremely important to start having discussions about just how differently we are loading our body for propriety's (or vanity's) sake. I don't recommend that you chuck your bra or underwear immediately; your support systems have not experienced the full weight of the bits that need supporting in some time. However, I do suggest that it couldn't hurt to take a good hard look at your list of issues and compare them to your loads, and then take a stepwise approach to

NATURAL LOADS AND CANCER

While regular exercise is a commonly cited preventive measure for many cancers, specific loads have yet to be evaluated directly in cancer research (although non-specific ones have had a positive impact on reversing cancer cells; check out the American Society for Cell Biology reference for more info). Research into the mechanical environment of cancer cells is new, but at this time there is still no research being done on the loads we create (naturally and unnaturally) and the environments that are more prone for tumor development.

Inflammation is a better-known "environment" that fuels the cancerous flame, but what's not as well understood is the role mechanosensing plays in inflammation. With time, this will all become more clear.

reducing support over a long time, giving your tissues a chance to strengthen over time.

MOVEMENT MATTERS

It's clear that there is a major mismatch between the loads we make in modern life (sleeping in our beds, driving our cars to work, sitting in front of our computers, and vigorously exercising for sixty minutes a day, then sitting in front of the TV, repeat, repeat, repeat) and the loads we would have made (searching for, gathering, and preparing our own food, walking for water and building materials, carrying our home and children in our arms, repeat, repeat, repeat) were we living more in nature. No, this is not the point where I tell you that the solution is getting rid of all your clothes and moving into a cave. The solution will be much simpler than you realize. The difference between *you in nature* and *you right now, reading this book* is so great that even tiny adjustments to your loading habits can be worth millions in unspent healthcare dollars and bring about tremendous relief from your load-induced ailments.

If you want your health to change, you must change the way you move, and the way you think about movement. So how should we think about movement and our movement habits, and how can we change them?

A MOVEMENT STUDY OF THE HADZA

Through observation it's clear that hunter-gatherers have to move a lot, but in an attempt to support the physical activity component of the mismatch hypothesis, scientists wanted to quantify both cardiovascular disease risk factors and the daily movement habits of hunter-gatherer subjects. So they asked forty-six Hadza (hunter-gatherers from Northern Tanzania who subsist on 90 percent wild resources) to wear heart-rate monitors for four two-week periods during different seasons, to quantify the number of minutes they spent in light, moderate, and vigorous activity (40–54 percent, 55–69 percent, and 70–89 percent of their maximum heart rate) each day.

They found that, on average, these Hadza spent 221 minutes in light activity, 115

CONT'D ON NEXT PAGE

CONT'D FROM PREVIOUS PAGE

in moderate, and 20 minutes in vigorous activity each day (with Hadza women engaging in 40 more minutes of moderate to vigorous activity than men each day) and demonstrated no risk factors for cardiovascular disease.

I found two additional tidbits in this study to bust our cultural narratives about movement. First, while Hadza men tended to walk farther each day, the women got their increase in heart rate from food-related movements (pounding nuts and digging) as well as carrying heavy loads (children, water, wood, and food) while walking. Second, the data did not show a significant decline in the amount of daily activity in Hadza between ages 18 and 60, as we often see in our own culture; instead, that age group showed an *increase* in moderate and vigorous activity. If you've been using age ("I'm too old to be moving around so much!") or chores ("I've got too many things to do and no time for exercise!") as a justification for not moving more, it might be a time for a re-frame.

Check out the Raichlen reference for more info.

THE DIFFERENCE BETWEEN EXERCISE AND MOVEMENT
CHAPTER 3

In other words, our bodies are complex hierarchical structures, and hence mechanical deformation of whole tissues results in coordinated structural rearrangements on many different size scales. DONALD INGBER

I have had the unique experience of being a couch potato (ages five to eighteen), an exercise addict (ages nineteen to thirty), and an all-day "natural" mover (ages thirty-one to thirty-eight, and counting).

I was extremely bookish as a child. I played and rode my bike like all kids, but if I look at the actual frequency of my movement in each twenty-four-hour period, I spent most of the time sitting and reading. I was a swimmer in high school (which required a couple of hours of swimming a day for a few months a year), but even then, practices were sandwiched between sitting at my desk or lying on my bed, reading. It will come as no surprise to you that my eyeglass lenses are as thick as a wall.

During my senior year I started walking the couple of miles to school, which made me feel so good I found myself driving to places where I could walk after I walked home. I started losing weight. My friends at the ice cream shop where I worked were in their twenties and more prone to gaining weight from the common four-scoops-per-shift habit (oh, to be a teen again). They joined a gym and I came along.

This is where I met the StairMaster. And the aerobics classes. And the squat rack. This is where I triumphantly ran my first mile by choice (I sucked at running the mile every year before that) and where I fell in love with the exercise high.

In college I was still bookish and loved ice cream, but I found I could manage my weight by running hard every day. At first it was two miles, then five, then ten. When I left the physics department to study biomechanics in the kinesiology department, working out (for credit) was required. One semester I took a running class, an aerobics class, and Kinesiology 20—a thrice-weekly exercise class required of all kinesiology students for the purpose of passing the required fitness testing. *In one semester* I went from running a 10-minute mile to running a 6:45-minute mile. Holy crap, right?

While in college, I decided that it would be best if I could get paid to exercise. This would make sure that I could pay for college and stay in shape all at the same time. I took my group exercise instructor certification and added ten classes a week to my already scheduled workouts. I ran, swam, strength-trained, and cross-trained with free weights and body resistance. I taught stretching, core strengthening, kickboxing…you name it, I taught it.

I did this for almost ten years and was in fantastic shape. Except my body hurt all of the time. I had horrible acne on my jawline. And the worst thing was, I started to panic if I couldn't get my exercise session in.

My other exercise friends were the same way. We were all capable of huge cardiovascular endeavors (like running four miles at five a.m. and then popping in to teach two back-to-back cardio classes where we had to talk the entire time), but we weren't really strong. Almost no one could do a pull-up and many needed some sort of brace on to perform physically. Most were not lean and many had pretty high body fat. My menstrual cramps were horrible, and then there was the time I threw my back out…sliding an ottoman across the living room. While

I certainly won the respect of my friends and the title of "healthiest" member in my family, I wasn't really healthy—at least, I didn't have the level of health I desired. Despite my regular exercise and adhering to all the protocols found in an "evidence-based" program, I didn't feel good.

It wasn't until I went to graduate school and studied biomechanics on a cellular level that I understood why, and ditched my exercise program for a movement lifestyle.

And boy, there is a huge difference between movement and exercise.

MOVEMENT AND EXERCISE: THE DETAILS

If a picture is worth a thousand words, then surely a Venn diagram is worth two hundred and fifty thousand. Which makes this section worth a million words, so I hope you're ready.

Here is a diagram titled Movement.

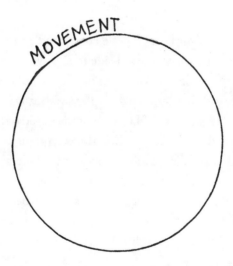

Right now the diagram is empty, but you can list anything and everything the human body can do in this circle. Snapping your fingers, bicycling, unicycling, squatting, winking, breathing, burping, slacklining, breastfeeding, birthing a baby, waving your arms around, lifting, diving, picking apples, and even something as tiny as getting goosebumps could all be accurately placed in this Movement diagram. See?

You have moved today. Your heart muscle contracted and released. Your lungs inflated. Your eyeballs might have moved side to side, vigorously consuming this book. You probably got out of bed, sat down on the toilet, used your hand, elbow, and shoulder to wipe, and then stood back up again. But if you haven't exercised, there is a chance that if I asked you "Have you moved today?" you might answer "No, I haven't moved today. Not really." Because even though we understand that of course there are thousands of possible movements made by the body, when I talk about *movement,* most people think *exercise.*

Exercise is the healthiest of all movement, or so we've been told. Ask any expert and they will probably say that, of course, you cannot be healthy unless you exercise, right? So when it comes to movement that *really matters,* it always boils down to exercise.

Just thinking of the term *exercise* is likely to cause your brain to respond by conjuring up mental images of fitness machines, athletics, yoga class, pull-ups, dance class, lifting weights, running, pedometers, and heart-rate monitors. In short, your mind currently organizes movement information like this:

I'm going to stop you here because the *first step* to radically improving your health is to let go of the notion that movement is exercise. To move your health forward with movement, it is essential to mentally rearrange the relationship between movement and exercise in your mind, so it looks more like this:

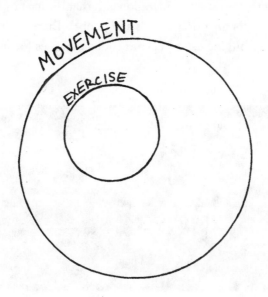

I want you to keep exercise and movement separate in your mind because there are many movements we wouldn't consider exercise that are essential to the tissues of the body. For example, the workings of an infant's mouth while feeding at the breast are different from the workings of an infant's mouth while feeding off a bottle. In the end, the task of getting milk is accomplished no matter if you take a boob or a bottle, but the *process* of milking the breast, it turns out, is important to the optimal formation of the jaw and face bones. The structure of the face bones and established motor patterns of the face muscles end up affecting other processes, like breathing and swallowing, as well as the space available for tooth eruption.

Breastfeeding is a human movement that results in a particular bone robusticity the body calls on in the future, yet most people would not place breastfeeding into the movement category (because they don't have a movement category; they only have one for exercise). Most people would not think that a movement as subtle as breastfeeding—a specific tongue action done for a dozen or so minutes at a time throughout the day, when you were an infant, no less—would have an impact on the health of an adult.

This is why it is important to start thinking about non-exercise movements. There are too many essential movements and loads created through movements—subtle movements that escape consideration and that most of us are missing (like the swinging of the aforementioned balls and boobs). There are entire categories of movement missing from the list of recommended exercises for "health."

ANATOMY BIT

To understand stress risers, imagine that you install a titanium towel rack by screwing it into your bathroom wall made of adobe mud. With the regular application of force (in this case, placing your towel on the rack day after day), the strong rod would cause damage to the weaker wall due to the disproportionate strength between them. Even though all cells are cells (deep, I know), the collective strength of a group of cells can mean one area of a tissue is mechanically different from another just a few cells away. The key to keeping all areas of your body strong is to use your body more evenly.

WE NEED TO CROSS-TRAIN, KIND OF

If you've been an exerciser for some time, then you've probably heard the term cross-training. Cross-training is an exercise technique that calls for a variance of exercise type (and sometimes a variance of the way you do one type of activity) to improve the body's response to exercise and decrease injury.

Cross-training is often recommended by exercise and therapeutic professionals because being strong or healthy in one specific way doesn't transfer over to being strong or healthy in every way. For example, if you're a kick-ass runner and you run all the time and you run easily and without pain, your body is probably very adapted to the tissue strengths necessary to run. But say you help a friend move and, in lifting a few boxes, you are thrown out of commission for a month with a major back spasm. Weren't you strong? Weren't you in shape? Weren't you a regular mover, and doesn't that mean you were conditioned to move? The answer to all of these questions has to do with the Law of Specificity, which dictates that you get better at what you do and not better at what you don't. A particular way of moving creates loads and adaptations in *only* the tissues pulled and pushed and compressed by that activity.

The more you do the same thing over and over again, the stronger *some* areas of your body become. As with any kind of material—cloth, metal, wood—strength in one area increases the relative weakness of the surrounding areas *not* strengthened. Having strong, regularly used parts next to underused (or overused) weak ones can actually increase tissue damage by creating a natural stress riser.

The more parts you integrate into your movement, the less disparity between parts. This is why scientists have recommended changing up your workout (i.e., cross-training) for years: to close the strength gaps that live between the exercises you do and exercises you don't.

The gap between what you are physically doing right now and what your body is capable of doing is vast. Typically, cross-training is applied to your workout: "If you usually run, you should throw in some strength training," or "If you are a cyclist, you should balance your workout with yoga." Again, the sentiment is correct. Look at what you are doing and then round out the loads applied to the body by doing something different. The problem is, most of what we do for exercise is just slightly different versions of the same kinds of motions. There is rarely uniqueness to what we do, and even routines designed to recruit more of our bodies often fail to do so because our movement patterns are so ingrained that we don't perform the program's movements as intended. Moving differently requires an inordinate amount of mindfulness, and once we forget to pay attention, we automatically default to the same pattern and go back to increasing the difference between one and every other area of our body. We go back to making stress risers.

In order to determine where we move a lot and where we move a little, we must look beyond our exercise session to the frequencies of movement, ranges of motion, and the way we load our body throughout the day, every day, over a lifetime. And instead of balancing our workouts over the week with different modes of exercise, we need to "cross-train" over the day—so that our workout isn't the only peak on the "how much did I move today" graph, and the bulk of our non-exercise time isn't shaping our tissues into the same use patterns over and over again.

If changing the way we load our body over a few workouts a week can help us avoid injury by reducing stress risers, how about changing the way we load our body over an entire lifetime?

THE DIFFERENCE BETWEEN MOVEMENT AND EXERCISE

There are a few reasons I've drawn the Exercise circle so much smaller than the Movement one. The variable of time is perhaps the easiest place to start. Clearly, the amount of time we allot to "exercise" is very small compared to the amount of time we are *capable* of moving our bodies, which is 100 percent of our waking hours.

But this graphic statement doesn't only show the large difference in time spent on exercise rather than movement. The relationship between movement and exercise has even more to do with the geometrical shape your body assumes when you exercise. Most exercise consists of the same movement or limited number of movements, repeated over and over again. Say you ride a bike or run for cardio an hour every day. If I were to quantify what your joints are doing for that hour, I would find that you are cycling through a very small set of possible joint ranges of motion over and over again. Even modalities of exercise like a yoga or dance class, which focus on cycling through twenty to a hundred unique body positions, offer a small fraction of the movements we are capable of.

In *Simple Steps to Foot Pain Relief*, I write about evaluating the crazy number of unique positions the foot is capable of. Because each foot has 33 joints, the number of unique ways the foot can be distorted is enormous—8,589,934,592 (found by calculating 2^{33}). It's also important to mention (as I failed to do in my first book) that to make this calculation, I had to assume each joint could only move to one other position, which is like saying your knee can be either all the way bent or all the way straight. If we allowed three positions for each joint, the number of unique foot positions would be 3^{33} (about 5.56×10^{15}, which is 5,560,000,000,000,000), four unique positions 4^{33} (about 7.38×10^{19}, which is 73,800,000,000,000,000,000), and so on.

Now extrapolate to the rest of your body, which has over 300 joints. This means that the number of unique positions possible by you (if we only allow two positions for each joint) is 2^{300}—which, calculated, is 2.04×10^{90}. I'd love to remove the decimal and add the eighty-eight zeros for effect, but instead I'll just write this: In light of the body's capacity for this almost incomprehensible number of geometrical positions, cross-training as we do it—adding some yoga to your regular running routine or swapping a dance class for a cycling class—seems like a pretty small step toward moving more.

I mean, it's a step, for sure, but the notion of cross-training as used in exercise is the smallest possible way to think about the largest thing you can do. (And if I'm doing my job right so far, then you are considering what a small range of loads your body has experienced this last decade, this past year, and since you started reading this chapter. Feel free to change your reading position.)

UNNATURALLY MOTIVATED

The final difference between movement and exercise doesn't really fit well into a Venn diagram, and it boils down to the brain's precursor to movement: motivation.

Although Arthur C. Clarke's *2001* is in no way an anthropological textbook, the first chapter describing the "thought" process of pre-human animals is my most favorite passage of any book describing prehistoric life, ever. In *2001*'s first chapter—before the aliens come—Moonwatcher, the lead primal "person," is motivated only by his hunger. Hunger initiates movement. Movement is his only answer to the unformed but literally gut-wrenching question, "Where's my food?"

Now imagine a different scenario. Stuffed on the couch after a holiday meal. Pants straining at a restaurant. All of that readily available food—brought right to your table—no movement required. And then there are the thoughts! "Oh man, I ate so much I feel sick!" "I need to hit the gym!" "I need to work this off." Or how about answering this question: How many times have you justified eating more than you needed—more than you even wanted, maybe—by saying, "I'll work this off later!"?

In our lifetime, it is the abundance of food intake that motivates us to exercise. When it comes to both food and exercise our relationship is not only off, we are essentially operating in *reverse* when it comes to our human reflexes to eat and move.

The current construct of abundance with respect to foodstuffs affects the way we think about and initiate movement programs. The ancestral model of movement says movement initiates reflexively via a desire to find food. In order to satiate biological hunger, movement was initiated in an organic way. When food became readily available without movement, our relationship with food and movement completely changed. Now we move as a response to too much food. This perspective, that movement is necessary to mitigate the effects of food, is a mantra repeated by every health publication and practitioner. The notion that movement's purpose is to avoid the negative repercussions of consuming food is fundamental to our modern beliefs about health. Given research on the power of how we think about exercise and how well we adhere to a plan, it's important to note that we've essentially reversed the natural thought processes around both movement and food.

Our current model sets both food and movement as a negative. Eating food (a biological imperative) makes us feel guilty, and we turn movement (also a biological imperative) into punishment. Atonement for diet through exercise. It is no wonder so many people feel unable to begin moving (and eating) in a way that honors both food and our innate ability to move. We've got it all backwards in our minds.

I was lucky enough to have an entire chapter, but if I only had one paragraph to state the most fundamental difference between exercise and movement it would be this: If the goal of exercise is to reap the physical benefits of movement, the goal of non-exercise movement would be to reap the non-movement benefits of the activity. Going for a one-mile or thirty-minute walk to strengthen your legs, burn some calories, and stretch your muscles is an example of exercise. Walking a mile to the store because you need to pick up something for dinner is an example of movement. Both may use the body in exactly the same way, but there is a difference in the bigger picture regarding how we think about and schedule the needs of our body. As you read more of this book, it will help to remember that *exercise is movement, but movement is not always exercise.*

SO YOU ALREADY EXERCISE, GREAT!

If you already exercise, that's awesome. I love to exercise too. Clearly exercise can and does result in many positive outcomes. Decreases in body fat or increases in muscle, better endurance, improvements in functional strength, and better joint ranges of motion are just a small sample of some of exercise's benefits. There are also reasons to exercise that go beyond the physical reward—competitive athletes exercise for income or status. A dance class can result in joyful self-expression. Some professions require exercise in order to complete job-related tasks. Research shows that many people in their golden years exercise not for fitness benefits but for company. Exercise can bring people together or create a temporary escape. Exercise makes us feel good. Exercise makes us feel better, both physically and mentally.

But here's the thing: Exercise does not always make *every part* of us better. What can be good for the mind might be hard on the knees. What can be nice for the waistline can cause your pelvic floor to fail. What can serve as a mental escape now can cause mental anguish down the road. We can do better than exercise. We *need* to do *more* than exercise. Our bodies require more "movement nutrition" than exercise, as we do it, provides.

Exercisers represent the movers in our culture, but exercisers themselves are sedentary most of the day when compared to hunter-gathering populations. The difference between a non-exerciser and someone who works out regularly is about three hundred minutes a week. For those who have ever started an exercise program, those extra three hundred minutes can be tricky to commit to, but when you do, you feel so much better. Imagine the difference between the average exerciser (moving about three hundred minutes a week) and the hunter-gatherer, who used his or her body eight hours a day (approximately three thousand minutes a week), and still took more time to rest than you do. The frequency the hunter-gatherer moved is tenfold to our movement frequency. The difference between three hundred and three thousand minutes of weekly loading is tremendous for all of our tissues.

In order to improve our health we must recognize the limitations of the exercise model. To again use food as an analogy, our current exercise model calls for very few "total calories" a day. The government currently recommends three categories of movement—cardio, strength training, and stretching—which can be likened to the macronutrient groups carbohydrates, fat, and protein. Without a better prescription for what to eat/how to move, the moves we consume to meet our recommended daily allowance (RDA) become mostly junk. And, without considering how each movement results in a unique load profile and a specific adaptation, we will miss how exercise in a modern context can actually *create disease*, despite our very best intentions.

THE HEART OF THE MATTER: WHY WE MAY NOT NEED "CARDIO" AFTER ALL
CHAPTER 4

In his book *Nature's Garden: A Guide to Identifying, Harvesting, and Preparing Edible Wild Plants*, Samuel Thayer, wild food forager and naturalist, lays out just how ridiculous it is to not *value* the seemingly endless details found in a biological system. In fact, he did this so well, I asked his permission to include a paragraph from his book here:

> Whoever first said "The Devil is in the details" must not have liked details. And I doubt that he was an economic botanist. Because when it comes to edible wild plants, the *miracle* is in the details. It is the details that give one the power not only to identify plants, but also to select the best specimens among them....So do not shy away from details: and don't resent Nature for being so replete with complexity. That is its glory, not its downfall. We owe our very intelligence to this miraculous complexity. It is not the burden of the naturalist to learn this complexity; it is the awesome reality. More than anything else, which of these attitudes you choose will determine your success. So learn your details with pride, and experience them with gratitude. Let the details excite you—for there are enough of them to excite you for the rest of your life.

Now to be clear, I don't know anything about wild plants. (Or I didn't, until I read his book. I have enjoyed preparing dozens of wild foods this last year thanks to Mr. Thayer.) What I love about this paragraph, and why I wanted to include it, is that the sentiment applies to every application of biological principles, and it's a good preparation for what comes next: an in-depth look at our cardiovascular system, and all the ways in which our paradigm of exercise is lacking in detail.

HEMODYNAMICS AND EXERCISE

Most people know that the cardiovascular system delivers oxygen and nutrients to the body while removing waste from the tissues. They believe this process is driven by the heart muscle, using the arterial and venous system to circulate the blood. This definition is not exactly incorrect, but it is oversimplified to the point of being misleading when used in the context of improving cardiovascular health.

If I had you create a mental image of the cardiovascular system, chances are you would come up with an image similar to those presented in anatomy and physiology textbooks throughout the world:

But this picture is incomplete. The arteries and veins are usually the main feature of drawings like these, but what this image leaves off are all the smaller vessels where oxygen delivery—the very reason the blood needs to circulate through the body—takes place. The final destination of oxygen is capillary beds—the teeny-tiny tubes that branch off the arterioles, which branch off the arteries. This is where the exchange between the body and blood occurs.

If you prefer pictures to words, it means that the blood flow to the hand pictured as this:

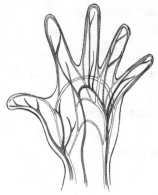

really looks more like this:

Even this picture hardly does justice to the dense network of your capillary system. The capillary system is so prolific, most cells in your body (and there are a hundred trillion of them) are within fifty micrometers of a capillary. Don't have your micrometer measurer handy? A human hair is about seventeen micrometers thick, which means almost every cell in your body is within a few hair-widths' distance from a capillary.

ARTERIES ARE ONLY THE HIGHWAYS TRAVELED BY OXYGEN

Say you're trying to visit a friend who lives a few states over from you. You hop in the car and head over on the fastest freeway to get as close to her house as you can before jumping onto a smaller but still crowded inner-city road, then to the slower routes through suburbia, and finally pull your car into the driveway of your friend's home.

The process of going from being one in a thousand cars driving on mega-highway to being the only car in a driveway is similar to the experience of a red blood cell traveling from your heart to a capillary bed. The function of the cardiovascular system is only "good" if it accomplishes the task of delivering oxygen everywhere in the body, and the state of your cells—your *microhealth*, if you will—is a better indication of the function of your cardiovascular system than is the ability to run five miles.

HOW DOES BLOOD-OXYGEN GET OUT OF THE ARTERIES AND INTO THE CAPILLARIES?

Blood moves like this: The mechanical stimulation of a muscle working causes the smooth-muscle walls of the arterioles to relax and open (this is called vasodilation), causing a drop in pressure that pulls blood from the arteries to the capillaries.

Now I'm going to stop you here and have you recall the version of the cardiovascular system you are most familiar with. Do you remember any mention of the musculoskeletal system? Probably not. The cardiovascular system most of us are presented with is one in which the heart is essentially pushing the blood around your body. In reality, working muscles pull your blood to the tissues that need it. The "heart-pump" model of the circulatory system is probably why the "strengthening your heart" paradigm has persisted. Within a sedentary culture, the heart becomes the sole mover of blood. This is not "how the body works," but how the body operates in a movement drought.

The total volume of the capillary system is roughly equal to the volume of blood your body holds—about five liters. Which means that if your entire volume of blood were to move from the arteries to the capillaries there would be little left for the heart and major vessels. Oxygen delivery is extremely important for you and your cells to survive. Your cells, without regular oxygen, react much like you would without regular oxygen. Pretty panicky, right? Yet if you had enough blood to fill all your blood vessels all the time, you would be very heavy and inefficient. In this case, the trade-off your body has made for constantly saturated oxygen is lightness.

Now let's think about that whale in captivity again. It makes sense that the structure and functions of an animal adapt to the movement environment it has experienced while evolving. Just as the whale fins depend on the mechanoenvironment created through natural swimming, our very own system of oxygen distribution depends on frequent and constantly varying muscle use.

Because the body's oxygen delivery system is based on use, we get the best of both worlds: constantly available cellular nutrition when we need it, without the burden of extra weight. The only caveat to this arrangement (always read the fine print!) is that the displacement of blood into the capillaries depends on the use of the musculoskeletal system. If you don't move, your cells don't get fed. If your cells don't get fed, they die. So now we have yet another reason movement is not optional: In addition to creating loads and modifying genetic behavior, movement is an essential step in the process of oxygen delivery.

WORKING FOR A LIVING

As we've already learned, each mode of exercise uses particular muscle groups in specific ways. Working muscle groups need more fuel, create more waste, and experience an increase in blood flow. But regular exercise does not automatically imply that all your parts have received nourishment as a result of the time you spent moving; the benefits of exercise apply only to the areas that are working.

This brings me to something I'd like to clear up with respect to working out: *The effect of exercise—specifically the increase of oxygen delivery—is not systemic.*

Our use-based cardiovascular system means that blood is constantly being shuttled from working place to working place, feeding and removing the waste of participating tissues. Just like an airplane, the more your body works, the more fuel it requires and exhaust it creates. But to go a bit deeper, your body is not really the airplane in this analogy; each muscle is itself the airplane.

You've got six hundred skeletal muscles in your body (give or take) and there is a good chance that your exercise program does not use all of them.

STANDING IS THE NEW SITTING

And then there's the whole frequency thing that needs addressing. By now you understand that a daily bout of activity swaddled in sloth means you're missing out on essential loads. So that bout of activity doesn't decrease your risk for disease. And in fact, recently, researchers have discovered that *sitting time itself* is a risk factor for cardiovascular disease, even in those who exercise regularly. Regular bouts of exercise do not undo the effect that sitting has on the body. As you can imagine, this was reported widely under headlines like "Sitting Is the New Smoking."

I know, it seems like "sitting" is referring to that position you adopt in a chair or a couch, but consider this: If I told you that sitting kills, then you'd swap standing for sitting, naturally thinking you'd avoided the problem. But if you just stand there in your office for fifteen years, then you'll likely end up with issues equal in quantity to those of the guy who sat all day. Let us not forget that the collective move of workers to chairs stemmed from the standing injuries created by post-industrial standing-all-day factory work. "Standing" already has its risk of injury on file.

The move from your work chair to a standing workstation is like that joke, "I read that all accidents happened within fifteen miles of one's house, so I moved." "I read that sitting kills, so now I'm afraid to stop standing." For load reasons, the transition from a chair to a standing workstation is a step in the right direction, but standing the bulk of the day is a different version of the same problem. The position of sitting isn't problematic, it's the repetitive use of a single position that makes us ill. (Okay, the geometrical position of sitting is actually a *separate* risk factor for cardiovascular disease, but I'll get to that in a bit.)

Imagine if the SITTING KILLS headline were replaced with BEING STATIONARY SO MUCH OF THE TIME KILLS. Would you be motivated to stand more, or motivated to move more? (I'd probably want to change the word *kill*, because there's no reason to be a downer and we all have to die sometime. In fact, I think that any health initiative marketed as "Do X to avoid death" is also missing the point. There is something between living and dying we are after, and it has to do with biological success—being able to perform basic human tasks without medical assistance—and feeling good as a holistic being.)

So. The circulatory system works all the time, whether your muscles help or not. Taking your muscles out of the equation basically forces your heart to do all the work, all day. Then, when you're ready to stand up, you ask your heart—which has had to pick up your muscular slack all day—to engage in an intense activity for the purpose of keeping it strong enough to, again, do all the work.

BLOOD VESSEL GEOMETRY AND BLOOD FLOW INJURY

I've heard disease—like an osteoarthritic knee, for example—casually explained away with statements like "your knee just got old and wore out." Why, then (in this case), is the other knee fine? Aren't they the same age, after all? In the same way, people are always surprised to learn that if they do have plaque in their arteries, they don't have it all over. So, why one place and not the other?

The hardening of the arteries in response to the accumulation of plaque is called *atherosclerosis*, but plaque accumulation is not systemic. There are certain locations where plaque accumulates. There are some arteries more prone to plaque than others, which should be an indicator that diet isn't the only place we should be looking when trying to prevent plaque accumulation. I mean, diet is super-duper important, but the prevailing theory in the biomechanical community is that plaque accumulates in areas (the abdominal aorta, iliacs, coronaries, femorals, popliteals, carotids, and cerebrals) where the patterns of blood flow are the most complex.

The behavior of blood is different depending on—you guessed it—its mechanical environment. Just like water can be an easygoing stream or a crashing wave beating at the coastline, blood can be benign or not, depending on the forces placed upon it.

The areas of the body most prone to plaque are also the areas with the most complex vessel geometry. Areas where arterial tubing isn't straight—where arteries branch, loop around, or pass over a major joint—are where you'll find plaque most of the time. Research shows that slow or varied (oscillating) flow patterns create a "non-aligned" shear stress as the blood moves along the wall of the blood vessels. Instead of blood moving *along* the endothelial surface, blood

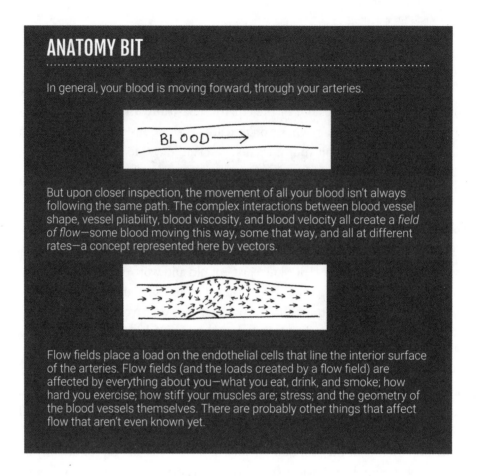

flows *into* the surface. Over time, this low-grade, repetitive mechanical interaction between blood and the lumen's epithelial lining causes the genetics of these endothelial cells to change the way they express themselves. Cells that line the arteries are, by nature, atheroprotective (they protect against the formation of plaque), but after repetitive mechanical strain created by malaligned blood flow, these cells become atherogenic—which means they begin *promoting* the formation of fatty plaque.

I liken the load-via-blood-flow phenomenon to a potter throwing a vase on a pottery wheel. In this case, the lump of clay is a cluster of epithelial cells, and the way the cells are stimulated is the result of the pressure of the potter's hands, the position of the potter's hands, and the spinning rate of the pottery wheel. Each of these variables can change how the pot looks, right? Pressing

SHEAR STRESS AS APPLIED TO BLOOD FLOW

The tangential force of the blood flowing along the endothelial surface of a blood vessel.

a little harder, changing the wrist angle, or slowing the wheel way down all change the interaction between the potter and the pot. Similarly, the habits you have—movement and position included—shape the flow patterns of your blood, which in turn shapes the cells, which in turn determines where and how much plaque you have.

Our point of view on arterial stiffening needs to be reframed. The hardening of arteries is seen as a disease—as "the problem"—but the endothelial walls' bulk-up-with-plaque response to mechanical strain could also been seen as an adaptation—a short-term improvement to the cellular lining. Without a thicker wall, the lumen could, theoretically, become porous, which would surely be more detrimental in the short-term for the survival of the body. So again, is our physiology broken, or is it responding completely appropriately to our particular (mis)use of our bodies?

LUMEN

In biology, the interior of a tubular structure.

In general, movement is beneficial in that it keeps the blood moving more uniformly. But picture this: all your blood, flying in formation (called laminar flow) through an open part of an artery, only to hit a narrow spot that came up quickly. What happens in this case is similar to what happens to water when you cover a hose with part of your thumb. It spurts! And goes from flowing smoothly to flowing turbulently (i.e., all over the place).

Those who have already begun to accumulate plaque, and those who already have accumulated risk factors for cardiovascular disease—the red flags that alert

ANATOMY BIT

Atherosclerotic plaque occurs preferentially in the abdominal aorta (down the center of the torso), iliacs (located on either side of the lower lumbar vertebrae), coronaries (on the heart), femorals (passing over the front hip), popliteals (passing behind the knee), carotids (running up both sides of the neck), and the cerebrals (in the brain).

you to problems on the cellular level—often prioritize cardiovascular exercise over everything else to address their "heart problems." But the problem is not with your heart. Your cardiovascular problems are arising from the sum total effects of everything you do, and cardio isn't the answer. In fact, cardio—a lot of it, consumed in isolation of all other movement—can be a turbulent-blood maker! Instead of short (relative to the entire day) bursts of high-intensity exercise, consider moving in a way that gets your blood moving (which is good) throughout the day—and massage your endothelial lining into a new shape.

Unfortunately, you aren't diagnosed as "having plaque" until you can easily see it, which means you have to have A LOT of it before it counts. Which brings us to this next section on the geometry of sitting.

As I mentioned earlier, sitting doesn't *only* create the problems that "being still" does, and here's why: Sitting makes your blood vessel geometry *even more complex* than it is when you stand. Yes, the loops and branches of blood vessels are there no matter how you are positioned, but bending the knees, flexing the hips, rounding the spine, and jutting the head add more bends into this system. If it were only a matter of minutes or hours, blood-wall interactions would not have time to accumulate. But when we're talking decades of the exact same complex geometry, the wounding adds up.

Needless to say, I'm not a fan of sitting and less a fan of sitting in a chair. But it's not only for blood flow reasons. There are (gulp!) even more reasons about to be revealed!

SITTING: THE LIST OF PROBLEMS GROWS LONGER

Let's say you read on your favorite website that "Sitting is Associated with Mortalities of All Causes." You decide that you're going to swap out your chair and desk for a standing workstation, and also take little walking breaks to increase your movement throughout the day. Awesome. By making these simple changes, you've introduced new loads to your bones and added more movement to your day without having to quit your job or alienate your entire office. You've also been sitting less, which means, if the research into sitting is valid, you've decreased your risk for disease. Well done, you.

The problem with researching the effects of sitting is that it is difficult to separate the *inactivity* created by the frequency of a single position from the impact a *repetitive geometry* has on the body. As an exercise scientist designing a movement protocol, the inactivity of sitting might be most prominent in my mind; inactivity means fewer calories are being expended, which, in turn, implies greater risk for obesity and poor glucose regulation—both risk factors for cardiovascular disease. So, as I've mentioned, sitting less and taking breaks are great ways of reducing the inactivity promoted by sitting.

A biomechanist, on the other hand, can have a different—or, let's say, an *additional*—take on sitting. A biomechanist could also consider: What are the forces created by sitting? How does the body adapt to long periods of continuous, similar input? What are the specific joint arrangements most common to sitting (90° hips and knees, a rounding of the lower spine, etc.) and what sorts of adaptations would you expect from these specific joint configurations?

If you've ever taken an anatomy course, you're probably familiar with the basic idea that muscles lengthen and shorten to move bones relative to joints. Say, for example, you sit down into a kitchen chair. To do so, your hips and knees all have to flex to about 90°. To accomplish this, some muscle fibers shorten and some lengthen, but, when you stand up, everything returns to its original, pre-sitting orientation. This is correct, in theory, but what isn't presented in this same anatomy class is what happens when skeletal muscle spends most of its time in a fixed position. Chronic positioning—that is, the frequent use of the same position, over and over again—makes the model of muscle movement more complicated.

Let's say you spend the hunk of the day in this position.

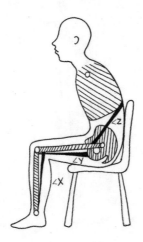

Or this one.

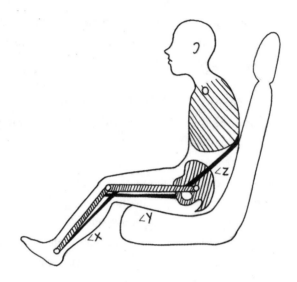

When you stand up, you probably think you are returning to this position.

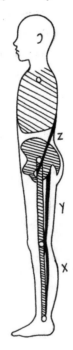

But actually, when you spend most of your time seated, and have since you were very young, when you get up from a chair your body doesn't straighten all the way back out. Your out-of-the-chair chair-lovin' body looks more like one of these people.

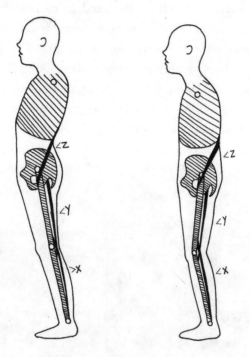

Most of us have assumed the sitting position through most of our lives, and in turn our bodies have adapted to sitting.

In a biological context, *adaptation* doesn't imply your body has been improved in the sense of having become healthier. Rather, adaptation is the result of your body's constant pursuit to conserve energy. Because we have practiced sitting daily, and for hours, our bodies have responded by making "sitting" easier on us.

Tissues that spend most of their time in a fixed position will adapt to that position by making alterations that are fairly permanent. The changes are not *truly* permanent, as they can change over a long period of time with new habitual behavior, but your tissues don't change as much as you might assume—certainly not just because you get up out of your chair at the end of the day.

YOUR CHAIR BAGGAGE

If you've ever had a body part broken and casted, you're familiar with the adaptations that come from immobilization. A decrease in movement is associated with decreases in muscle size (atrophy), vascularization (capillaries), the sensitivity in your proprioceptive system (your ability to sense where body parts are relative to each other), and bone mass. On the other hand, an under-moved area of the body will experience increases in the connective tissues found within muscle itself. I like to call these extra-connected areas your "sticky spots."

Being extra-connected is a great thing when you're talking about your family, your community, your career, or wi-fi. Having one muscle overly connected to another isn't so great. Immobility-induced connective tissue growth creates a sort of binding between muscle parts and behaves much like scar tissue. Instead of smooth, organized fibers, the chaotic arrangement of these asymmetrical fibers

ANATOMY BIT

Sarcomeres (Greek, with *sarco* meaning muscle, and *mer* meaning part) are the basic contractile units of skeletal muscle. Each sarcomere unit measures about 2.2 micrometers in length and, aligned in series, they form a myofibril (a thread that makes up a muscle fiber, which make up a muscle). The reason muscles generate force and move is that sarcomeres generate force and move. The net movement of a muscle is the sum total of tiny movements at the sarcomere level. For example, if 1,000 sarcomeres shorten by 0.1 micrometers, the entire series (read: muscle) shortens 100 micrometers.

Sarcomeres are made up of long, fiber-like proteins called actin and myosin that slide past each other when lengthening and shortening, and titin (the largest known protein)—which gives muscle its extensibility and elasticity. It is via the interaction of actin and myosin that the two halves of a sarcomere move closer to or away from each other in active motion, and it is via titin that the muscle "returns home" when finished working.

The force of a muscle contraction depends on the alignment of a sarcomere, specifically how the actin and myosin overlap at the time of muscle innervation. Sarcomeres with too much overlap (muscle that is shortened passively) have no space to move and therefore cannot contract and generate force, even if a signal to contract is sent via the brain. Muscle that is too long is created by a series of over-stretched sarcomeres, where actin and myosin have very little overlap and almost no ability to perform work. For a fabulous introduction to sarcomeres, watch the "Sarcomere length–tension relationship" lecture from the Khan Academy (see Resources and Further Reading for the link).

SARCOMERES AND MUSCLE LENGTH

A

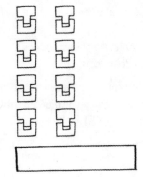

While muscles can work at many different lengths, sarcomeres cannot.

Here is a side view of the lower leg and the theoretical sarcomere arrangement for a "neutral" ankle. (Image A.)

Muscular models presented in textbooks are based on this sarcomere arrangement, where "in neutral," both sides of a joint have muscle lengths created by sarcomeres with optimal leverage—that is, the actin and myosin overlap is at a distance where a sarcomere has the spatial ability to get longer or shorter as needed. Neutral is neutral because, once moving, the sarcomere overlaps allow a muscle to produce force throughout the joint's full range of motion.

B

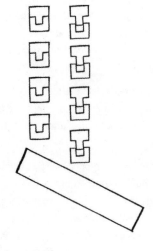

In this picture, the ankle has moved into plantarflexion (pointing your foot toward the ground), requiring the muscles (read: sarcomeres) on the back of the leg to shorten and the muscles on the front to lengthen. (Image B.)

Muscles change length because their sarcomeres do. Muscles become shorter when their sarcomeres increase their overlap; they become longer when their sarcomeres reduce overlap. Active moving requires the muscles (read: sarcomeres) on one side of a joint to get shorter while the opposing muscles get longer.

But here's the thing: The model here is the instantaneous rearrangement created by moving. Muscle behaves differently from what is presented here when you frequent the same position or small ranges of motion over and over again.

When you are in the same position chronically or use your muscles in a limited and repetitive way, your body responds to your preferences on the cellular level. Your body adapts to where you spend the most time, making it easier (i.e., taking less energy) to do what you do most often.

CONT'D ON NEXT PAGE

CONT'D FROM PREVIOUS PAGE

The problem with the sarcomere arrangement in picture B is that, once you've plantarflexed, neither side of the joint has any leverage left. On the back of the leg (the calf muscles), there is no physical space for the parts of the sarcomere to continue moving toward each other. On the front of the shin, the parts of the sarcomere no longer overlap. Without a physical connection, the sliding filaments cannot connect and generate much movement (or force).

Mechanosensors quickly "conclude" that the sarcomere arrangement in picture B is not advantageous in the long-term. If you can't generate force in a particular position, you are inert there. Sarcomeres pretty much have one way of organizing themselves, so if you, through your habits, change the average (read: most frequented) distance between muscle attachment points, your body will grow (in a process called *sarcomerogenesis*) or cannibalize (in a process called *sarcomerolysis)* individual sarcomeres in the effort to return the remaining sarcomeres to mechanical homeostasis.

At left, I've illustrated this adaptation by reducing the calf muscle's and increasing the shin muscle's number of sarcomeres by one. Thankfully, your body does this with a lot more finesse than I can recreate here, but this illustrates how now the sarcomeres are back to being able to produce force. (Image C, left.)

C

In order to move actively, sarcomeres must have the ability to shorten or lengthen, so it's really great that you can change yourself on a cellular level like this. But as with all adaptations, there is a cost.

One cost in this scenario is that the total producible force in image C is less than that in image A. But more importantly, when it comes to joint health and range of motions, the additional mass at the cellular level structurally maintains a plantarflexed position without any force. Not only can you not stop pointing your toes (or stop the hippity-hop "spring in your step" when you're walking), you can't dorsiflex your ankle, either. Have you ever heard of people who've worn heels so long that now they are physically unable to go without them?

Reductionist thinking about muscle has led us to conclude that we can simply stretch our way out of all the ways we immobilize our joints. Shoes, chairs, couches, kid devices, pillows, walking on flat ground. But each of these choices changes our makeup on the cellular level.

prevents the sliding action of muscle, limiting muscle force production and pain-free movement. On the cellular level, a sticky spot interferes with the transmission of forces throughout your tissue—mechanical signals that give cells context about loads placed upon them as well as their position.

LIMITED RANGE OF MOTION

Our "chair-shaped" sticky spots (and all the other repetitive-use spots we've developed) create problems when we load them with movement. When your body no longer articulates fully at a joint, you compensate by moving other joints instead. For example, your computer shoulders are tight and stiff. You take yourself out to play tennis and go for a sweet overhead serve by reaching your arm all the way over your head. Awesome. You thought your shoulders were too tight to reach up and overhead, but look at you! That racket is up there. Only, if we freeze that video (you had someone film your tennis game on your smartphone, didn't you?) we see that your arm doesn't go much above shoulder height, and when you hit the end of your shoulder's range of motion, your racquet traveled the rest of the way because you lifted the *rib cage* up to meet the racket. Long story short: You hit the ball, but you had to thrust your ribs (shearing the ligaments and compressing the discs in your spine) to hit the ball because your shoulders have adapted to computing and aren't fit for tennis at all.

Limited joint range of motion and a reduction in muscular force production can turn a movement session, done with the best of intentions, into an activity that comes with a heavy dose of damage. When we excitedly get up and out of the chair at the end of a long day at work to hit the trail or gym, we continue to retain a

FIXING YOUR STICKY

In addition to changing your habits and practicing your correctives, there are therapeutic modalities specific to reducing the adhesions between muscles. Fascia-specific therapies can accelerate progress when it comes to reducing sticky spots that interfere with joint mobility and muscle function. For a list of modalities (where you can find a practitioner) or products designed to "do it yourself," see Appendix.

FASCIA

A dense, fibrous connective tissue that permeates the body and forms a continuous, three-dimensional matrix that functions as a whole-body support system.

subtle chair shape. When you apply forces to a structure with a bony orientation and muscle leverages unsuitable for stable movement, you increase your risk of musculoskeletal issues. Your joints cannot bear loads optimally, the muscles are not able to respond to the load with much leverage, and your now-moving blood is bumping into bends that shouldn't continue to be there.

FORCE DAMPENING

Do you remember the T-shirt load lesson from chapter 1? Let's take that same shirt and spill something on it. Something really tacky, like oil-based paint. As it dries, it binds the fibers of the shirt to each other. What happens when you pull on that shirt now? The painted spot changes the way the shirt experiences the loads. A gentle tug on the shirt causes the areas *surrounding* the dried paint to stretch easily as they did before, but the area stiffened with the paint behaves differently. The same pull that used to deform the shirt down and inward now creates little (if any) distortion over the sticky area, while creating an extra-high distortion along the spot's edges. Just as sewing a patch onto a T-shirt would change the pull to the material, so does a sticky spot change a tissue's experience of load. Same application of force, different outcome—an outcome affected by the "sticky" relationship between fibers.

In this same way, the pushes and pulls of your body moving aren't trickling all the way down to every cell as they should. The cells in sticky areas of your body regenerate without movement context, or rather, a context that is stiff and unmoving; and the areas just outside of the sticky spot experience unnaturally high loads.

Over time, these sticky spots become cellular calluses and wreak havoc in the body—not just the effect of your movements, but also input from continuous forces, like gravity, which all cells should be free to experience.

I just called out the chair as a sticky-spot maker, but so much of what we do in our modern habitat is, essentially, immobilization. Shoes, desks, clothing, repetitive walking surfaces, repetitive actions required for any activity—these are all modern phenomena that cement, if you will, a narrow range of movement "into" the body.

AND NOW, WE MOVE

So. The largest environment still unexplored is your mechanical one. But the biggest limitation to this exploration will be the way we automatically associate movement with exercise. In addition to the psychological barrier the category of exercise creates, the physical limitation with exercise is that it cannot come close to replicating the variability and frequency of movements and loads necessary to keep our passive parts stable, our active muscles engaged, our blood flowing, and our sticky spots at bay. Exercise cannot come close to restoring the tissues already adapted to the way we have been using our habitat. In the same way supplements should not be the bulk of your diet, exercise should not be the bulk of your movement profile.

Loads created through movement need to be continuous and continuously varied to stave off adaptation to small ranges of motion. In short, exercise is good (certainly better than not) but it's not good enough. We need to think bigger than exercise, we need to think outside of the exercise box.

Ironically, you're about to enter the section of the book where I give you a ton of exercises. Because, well, the habitat of modern society automatically limits movement with the copious artificial terrain, man-made debris, and the bane of all human existence: the schedule. BUT, in addition to the exercises listed here—correctives, to decrease sticky spots and strength holes—are movement guidelines. Guidelines on how to move, sit, stand, rest, and even how to think throughout the day, that when followed, make every moment of your life count, on the cellular level.

MOVE

TRANSITIONING WELL
CHAPTER 5

In nature, there are neither rewards nor punishments—there are consequences.

ROBERT G. INGERSOLL

I am a superhuge fan of the minimal-footwear movement, but I am completely surprised at the failure of the barefoot/minimal-footwear community to regularly produce physiological guidelines on how to transition out of conventional footwear—as though the tissues in our feet were somehow different from those found in the rest of our bodies when it comes to adaptation.

When you decide you want to run a marathon, you don't jump off your couch and start with a twenty-mile run. I mean, most people wouldn't even consider jumping out of bed to sleep on the floor; they'd be so worried about their back or neck throbbing the next day. Yet people everywhere have gotten rid of their shoes despite decades of wearing them, and they've started running in minimal footwear without doing a lick of foot training.

Of course, many have transitioned just fine. Unfortunately, many haven't; even more unfortunately, it's only the injured folks who have taken their fractured bones and inflamed fascia to the podiatrist's office—reinforcing much of the medical community's data set with respect to minimal footwear and injury.

In the following chapters, I highlight missing behaviors like barefoot time, squatting, walking on natural terrain, and hanging and swinging from your arms. I'm going to explain how the mechanical environments created by these activities are essential, and I'm hoping to do such a good job explaining that you'll want to go and do all of these things immediately.

But I'm going to ask that you don't, and here's why: Your current body might not be able to adapt to major changes without injury. Or, maybe you can muscle through a lot of it, but generate a bunch of compensating mechanisms—mental programs that make it harder for you to find your reflexive strength.

There is a difference between saying, "It is natural behavior for a human to squat multiple times a day throughout their lifetime," and, "You need to start squatting ten times a day, starting now." My presentation of the details of ancestral human behavior is not a recommendation that you begin this behavior at once. After each section, I will list the transitioning behaviors and cautions for each category. Transitioning behaviors include mobilizing and strengthening exercises as well as the lifestyle modifications necessary to target specific muscle and joint actions. Cautions include looking for contraindications or modifications, and general guidelines on taking a stepwise approach.

I don't want to discourage you from working diligently; I only want to encourage smart training practices, which include small increases in loading behaviors in order to adapt correctly. If you throw down this book and head out to hike five miles of rugged terrain while wearing your five-fingered shoes and carrying a kid in one arm and a deer carcass on your shoulder, you're going to be hurting the next day, and maybe for the next few months. Extreme soreness has become a celebrated experience in our culture, but pain is often an indication that you've gone too far, too fast. By applying basic exercise-science principles, you should be able to improve your health while undoing old adaptations—without injury.

WALKING

Walking is the great adventure, the first meditation, a practice of heartiness and soul primary to humankind. Walking is the exact balance between spirit and humility. **GARY SNYDER**

Perhaps you're thinking, "I already walk, so what could I possibly learn here?" But walking is not walking is not walking. In the same way carrying thirteen pounds' worth of pumpkin can elicit different muscle use depending on *how* you carry them, the alignment of your body while you're walking can create different outcomes.

Just like an accent is the flair with which you speak, your *gait* is *the pattern by which you walk*. And the process by which you acquire your gait is very similar to that by which you acquire your accent. The way you walk has been influenced by many things along the way.

Gait-influencing variables include: your body size and shape; the way your family walked around you while you were learning to walk; your muscles' strength; your body's tension patterns; technologies (like diapers, footwear, chairs) used while you were developing your walking patterns; hobbies (like sport, dance, etc.); injuries; surfaces you walk on; and the surfaces you have walked on the most.

Because the shape of your bones and length of your tissues have adapted to your most frequently used positions and habits, and because most of the loads you have experienced are not similar to those experienced in nature, the way you walk right now is not likely to be your natural gait pattern.

Walking is often touted as one of the easiest and safest kinds of exercise a human can do. But most people walk so inefficiently that their very gait pattern is contributing to their spine, knee, or bone problems. Now, don't let this deter you from walking. If the choice is to walk or not, then by all means, walk. But by making just a few adjustments, you can change the way you walk so that you use more muscle, stabilize more joints, and create the necessary forces for cellular adaptation (like in bone, for example).

Walking creates a great portion of "what humans need" when it comes to load profiles, and even in the modern, non-exercising sedentary group, almost everyone walks—even if it is just to the car or around the house. *Everyone* who can physically walk can benefit from analyzing how they walk to determine their current strengths and weaknesses and work on improving areas where their "walking" is more like "falling."

ARE YOU WALKING OR FALLING?

For years, walking has been described as controlled falling. And I agree that describes how most people are walking around the planet. The muscular leverages created via chronic sitting (hip flexion) and positive-heeled footwear (plantarflexion) have robbed our muscles of the ability to produce enough force to facilitate stable movement. For most of us, moving forward via a rapid succession of falls is the only way we *can* walk. Had we been walking a lot more, sitting a lot less, and utilizing more of our bodies over a lifetime, our gait patterns would be entirely different. So although walking *can* be falling, it doesn't have to be.

When it comes to improving our load profiles through walking, we must assess the variables that affect loads while walking.

THE DISTANCE AND FREQUENCY OF YOUR WALKING

Duplicating the walking distance of hunter-gatherer populations is relatively simple. Anthropologists estimate that the daily tasks of gathering wood, food, and water, as well as migrating longer distances when necessary, meant that historical hunter-gatherer people were walking about one thousand miles a year. If we break this down into per-day mileage, we would have to walk 2.75 miles to match their distance.

But did hunter-gatherers really have "take a 2.75-mile walk" on their daily to-do list? Nope. Their yearly mileage was much more likely to come in shorter walks taken throughout the day sandwiched between days with very little walking and days with much longer mileage. The distribution of mileage, just like the distribution of thirteen pounds of pumpkin, matters. If you *always* walk 2.75 miles, you will not benefit from the physiological and tissue-strength adaptations necessary to walk longer distances. If you never bang out a ten-mile walk, then you won't develop the endurance to do so. Remember, the adaptations to even low-intensity endurance exercise sessions—adaptations such as increased capillaries and oxygen distribution—persist even when you're no longer participating in the endurance event. This means greater metabolic health (read: more energy expenditure) simply due to the state of your tissues.

So instead of dividing your twenty miles a week into tidy 2.75-mile daily walks, try another schedule, like this:

Monday: Two miles
Tuesday: Three miles
Wednesday: Eight miles
Thursday: One mile
Friday: Five miles
Saturday: Zero miles
Sunday: One mile

If you followed this plan, you'd create a different body than you would by walking three miles every day. You could also plan a greater physical endeavor, like a twenty-mile walk once a month or whenever your schedule allows. Our benchmark of a thousand miles a year is, of course, just an average. The climate and food availability for ancient hunter-gatherers must have fluctuated, and some years they likely walked more than a thousand miles, some years less. There is no perfect equation of movement behavior for us to duplicate; we simply need to increase our walking distances and their distribution to achieve a more robust adaptation.

In addition to considering how we distribute mileage throughout the week, we must also discuss how we distribute our walking throughout the day. As I've discussed, walking all at once brings mechanical stimulation all at once. Walking uses a greater number of muscles (when done naturally) than most other activities, which means taking yourself for a walk is like taking your cells out to eat. If you walk your daily three miles all at once, and then follow them up with stillness, your body must wait a full twenty-four hours until its next "feeding" and waste-removal session. If you walk one mile three times a day, the cells are fed smaller amounts throughout the day and waste is removed more frequently.

If we compare not walking to walking three miles a day all at once, then of course, walking the three miles is likely to leave you with better health. But remember that loads are affected by numerous factors, including the amount of time between load cycles. Being mindful about walking throughout the day can have a positive impact on your well-being, even if your "walks" are nothing more than one-to-five-minute laps around your home, office, or block. (Now quick. Put down this book and go walk around the block RIGHT NOW! Or better yet, go get the audiobook and listen to this entire book while walking.)

Seriously. Don't read any more unless you've walked for at least five minutes.

THE SURFACES YOU WALK UPON

We modern humans have not only spent most of our time in shoes (which create a repetitive environment for the foot), we have also spent most of our time walking over artificially level surfaces. Not just level surfaces, but flat ones. Which means our beautifully complex feet, ankles, knees, and hips have been casted, prevented from moving fully. By exposing these joints to the same surface, day after day, decade after decade, we have created structures that prevent the full use of our body and have lost the mobility necessary to cope with varying terrain. When we trip on a hole and sprain an ankle, we are quick to blame the hole. Stupid hole, in the middle of this playing field. How dare you appear right where I was walking!

A lifetime of picking the roads most traveled and the paths most groomed has basically ensured that our tissues are unable to carry our bodies safely through any detritus nature might throw in our path. The occasional hole is not the problem. Our weakness is.

As you design your walking program, remember the following:

- The more the foot can move and deform over a surface, the less the ankle is forced to do the work of the foot.
- The more inclined and declined your walking surface, the more variation in ankle, knee, and hip use, pelvic positioning, and the greater variance in muscles used throughout the entire body.
- The more the walking surface is cluttered with natural debris (sand, rocks, leaves, slate, roots, holes, pebbles, etc.), the more your feet, knees, hips, pelvis, and whole-body musculature have to work.

There are two general ways terrain can vary: *grade* (uphill, downhill, and how much of either) and *surface* (rough, slippery, bumpy, rocky, hole-y, etc.). Every unique combination of grade and surface results in a particular physical stimulation. When you compare the endless number of joint contortions and muscle contractions necessary for walking over natural terrain (think: hiking) to the single, repetitive pattern we use walking through a mall, it is clear, quantitatively, that the physical outcomes created by our current walking habits should be thought of as repetitive-use injuries.

"CROSS-TERRAINING"

If you've ever gone for a major hike, you know that your quads and hams can protest the next day. Changes in grade mean a lot more big-muscle use than you might be accustomed to. The loads associated with varying surfaces are typically smaller in magnitude, often requiring almost unnoticeable changes in foot bone or ankle positioning. Don't let the magnitude of the difference fool you—those differences add up.

Even though I'm a regular minimal-shoe-wearer, I, like you, log way more miles on groomed surfaces. Once on a road trip I stopped to meet my brother for a long hike to break up the car ride. We hiked for about five miles—a perfectly normal distance for me—but over slate. A surface I wasn't accustomed to. The next morning, my ankles could barely move. My body was totally adapted to minimal shoes, long hikes, uphills, and downhills. Unfortunately none of those are the same thing as working the muscles necessary for tiny adjustments in the ankle walking on slate.

I hope you can learn from my painful lesson. The small stuff is work. And it's not only the muscles that have to work more. These small deformations at your foundation translate into a constant stream of data to your brain and require constant communication throughout most of your body—the entire time you are walking. If you've been looking to improve your mind-body connection, walking off the beaten track is where it's at.

I did my own informal examination of the effect of walking surface on the body's cadence at a movement retreat in Hawaii. Research shows that walking "for fitness"—that is, using walking as a means to improve variables associated with longevity—requires you to walk *briskly*, which has been defined as taking a hundred steps per minute.

At the retreat, I had everyone test the rate of their normal walking speed in a paved lot where we were staying. (You can do this too. Have someone time you or use a stopwatch yourself while you count the number of steps you take in a minute.) Everyone found their regular strolling gait rate to be about 120 on average—well within the "walking for fitness" category.

After logging this data, I had them take off their shoes, and we timed them walking barefoot, to see if wearing shoes had affected their rates of walking. As you can imagine, everyone's pace slowed a little (even though this was a group of long-term minimal-footwear enthusiasts), as they were now a little more mindful about each step. Although the average steps-per-minute stayed above a hundred steps, it also dropped by about fifteen steps.

Next I had everyone move off of the pavement and into the Hawaiian jungle just outside the parking lot. With everyone's bare feet now facing tropical thorns, macadamia nuts, and lava rock, as well as terrain with endless holes and angles, we took a final sample. It's probably not surprising that gait rates slowed down to about seventy-five steps per minute.

Were the subjects working less hard walking barefoot in the jungle than they had been when they were walking in shoes in a parking lot? Of course not. They were working much harder, regardless of the number of steps they were taking per minute. So while we might say that the best way to walk for fitness is to walk at least a hundred steps a minute, we're really saying that this is the best rate for walking upon bland, unchallenging, and mind-unnecessary environments that cheat us of many of the benefits of walking.

In our quest to make everything safe, we've also made everything easy—which means that speed is the only variable we can use to make something challenging, which in turn limits the actual challenge of every walk we take. Walking on this long, boring path? Walk faster. That will make it better for you.

Walking quickly on a smooth path serves a purpose, sure, but when we only see speed (or intensity) as an indicator of body-use while walking, we lose the non-speed health improvements walking on varied terrain can give us.

CHANGE YOUR GRADE

The muscles and motor programs used in uphill and downhill walking are entirely different from each other, and both are different from patterns used for flat-ground walking. Most of us have spent most of our lives walking on flat ground, which has left our uphill and downhill muscles sorely underused.

Start looking at your walking plan and make sure you include hills. Does this mean you have to start looking for monster hikes? Of course not. Every grade—from the gentle slope of a friend's driveway to the hill at a local playground—has a nutrient you need. Adding big hills is great too, but if you think of hills in terms of percentage of incline, and walk as many different gradients as you can, you'll stumble on (no pun intended) the loads and muscular adaptation you've been missing.

LEARNING TO WALK TAKES TIME

Once, at the beginning of a six-month course, a student came up to me and asked if I could just tell her, real quick, everything she needed to know about walking so that she could practice it until the next session. I told her that, at the end of the six months, she might have a preliminary grasp of the big ideas that are necessary for understanding gait, and if she was really interested, she could spend a few decades at university studying walking at increasingly complex levels. Human movement is not a simple science.

Walking, as basic as it seems, is an extremely complex phenomenon that can simultaneously involve all your muscles at once. When I first started studying biomechanics, the lessons involving gait diagrams looked like the diagram below left. After a few more years, they started looking like the one below right (image adapted from work by Seireg and Arvikar, see Resources and Further Reading).

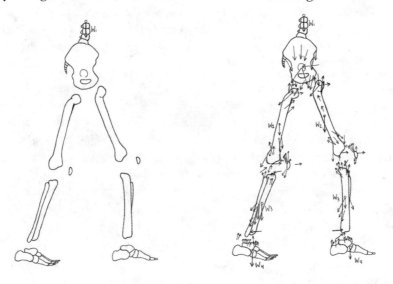

(Note that there are many contributing forces still missing! For example, there are no force arrows for the muscles of the feet, not to mention ones for entire tissue categories like skin and fascia.) Gait involves a coordinated effort between all your parts. For this reason, we will address gait at the end of the book—after you've had a chance to mobilize your feet, knees, hips, torso, spine, and shoulders. By working on these areas you will be, in fact, improving your gait—although this improvement is but a convenient side effect at first, with no need for you to make any conscious effort to change the way you walk.

A NEW VARIABLE: VARIABILITY

Restoring appropriate quantities of movement at each joint is something we modern movers must think about. One of the greatest variables of natural movement is *variability*. Environments that change constantly, in every possible way, ensure more thorough use of the body. Our body has become accustomed and adapted to repetition. At this stage in the game, moving like a hunter-gatherer requires more undoing than it requires doing. What you can undo, right now, is

---

PARADIGM SHIFT: RETHINKING THE DIRECTION OF RESEARCH AND THERAPY

The parameters for walking and gait research begin (probably unconsciously) with the premise that overground walking is flat-ground walking. Most anatomical models of the foot and ankle are based on the foot and ankle movements demonstrated by chronically shod, flat-walkin' folks. When we apply these models to human movement research without qualification, the therapeutic solutions to the many injuries and issues wrought by the diseases of behavior continue to evade us.

Generally, injury and therapy researchers miss (or at least fail to mention in their research) that level walking, performed at the frequency (or lack thereof) with which we have done it, is *in itself* a cause of injury. When we're seeking therapeutic solutions to chronic issues of the foot, ankle, knee, hips, and pelvis, we mustn't ignore the well-known impact that repetitive patterning has on tissue health. It would be a therapeutic failure if we did not address the root problem in the first place: that the way we walk (our gait pattern, routes of choice, and the frequency at which we locomote) is entirely unnatural for the human body. Which isn't to say that we don't need corrective exercise in addition to behavioral changes, but that a thrice-weekly set of thirty ankle swirls and twenty calf raises designed to "keep ankle sprains at bay" are hardly adequate to build the strength necessary for you to stumble into a hole and come out unscathed.

Thanks to the barefoot-running movement, some gait and anatomical scientists have begun to consider modern environments (like footwear or diapers in children learning to walk) and say, "Oh yes! Of course the impact of these items has to be considered when studying human movement." They are finally recognizing that what we know about how humans walk is actually based on how humans walk around in a modern context. If we don't acknowledge that our model of "normal" walking has nothing to do with nature, our therapeutic options are vastly limited. Imagine trying to solve a load-related issue like osteoarthritis in the knee without addressing the impact our walking environment and gait patterns have on loads. In order to optimize the health of our joints, we must utilize a wider breadth of their function.

---

your sitting habit. Adjusting a body part is the best place to start. Standing is another option. You'll get bonus points for sitting on the floor, placing your hips and knees in some unique position.

Anthropologist Dr. Gordon Hewes spent years gathering chair-free resting postures used around the world (you'll find these in the next chapter). You can quickly "paleo up" your day by selecting any of these poses instead of plopping into a chair. The more poses you cycle through, the better. It takes no additional time to add natural movement into your day, it only takes the breaking of the chair habit.

The same argument applies for positive-heeled footwear (a shoe with any sort of heel elevation as compared to the front of the foot) as well as any postural habit (chest up!) or gait pattern you've developed along your way. When you do something—anything—with high frequency, your tissues alter to the degree of needing a similar amount of time to undo by adapting to your new habit.

When you begin new movements, you have to slowly leave your old body behind to ensure the loads to your muscles, bones, and joints are what you want them to be. It's possible to make what appears to be all the physical movements of an ancestral population and not have a similar cellular experience because of your petrified bits. We are absolutely going to be delving into the larger, more challenging movements and feats of strength found in a natural setting. But alongside those will be measures and correctives that will help your body go back to the beginning—filling in those gaps of strength missing from your earliest years.

THE ALIGNMENT TOOLBOX

We know that we need oxygen to live; we know that tissues need to receive oxygen to be healthy; we know that the loads required to deliver that oxygen also mechanically stimulate our cells, another of their biological requirements. But how can we assess a movement's effect on the cells, to ensure that we are delivering the oxygen and mechanical stimulation they need? We need a tool to measure the loads, both on the whole body and on every body part. The tool I use is alignment.

Alignment is the study of the interrelationship between body parts, and between those parts and the ground. Alignment science considers how particular body positions change the various loads and forces generated within the body.

. . . .

GROUND REACTION FORCE

The ground's response to applied force; when you push against the ground, the ground applies a force back to you, equal in magnitude but opposite in direction.

. . . .

Changing the position of your pelvis to make your hip bones more or less weight-bearing, or lowering your rib cage to decrease the compression of the discs in your lumbar spine, are examples of adjustments to your alignment. Human alignment serves the same kind of purpose as car-wheel alignment: There is an orientation and appropriate range of motion to be found in every part that minimizes damage and promotes longevity of the vehicle—or of the body, in our case.

Correct wheel alignment for a vehicle is determined by factoring in the forces that are regularly applied to the car's wheels by the road, the weather, the style of driving, and the car itself. All of these forces are considered, as well as what the driver wants from the car (a race-winner, or tires that last a long time). Correct wheel alignment does not imply that the wheels need to stay in one position, of course; that would make driving awfully difficult. Rather, good wheel alignment allows the individual parts of the car the freedom to create the intended movements of the driver without causing damage to the car. When wheel alignment is "off," the behavior of one wheel can result in premature wear to itself or cause damage to the car elsewhere.

Movement is dynamic, with the number of forces varying at each bone and joint, and alignment is a tool used to measure these forces in a moment of time. By capturing multiple "moments in time," we can create a larger picture detailing the loads on a body over time. Human movement evaluation, on a larger scale, requires the consideration of all-day, every-day external forces (like gravity and the ground reaction force) and forces creating and created by movement. And, of course, all of these movements of the musculoskeletal system are resulting in forces on the cellular level, which impacts the state of the tissue that is comprised of cells.

Alignment in the context of our bodies is often confused with posture, but as with the car, correct human alignment does not imply that there is one body position that we should be using all the time. In fact, it is often our determination to maintain a "good" fixed posture that is undermining our health.

Alignment is a necessary tool when you're trying to quantify and compare historical and modern movement. Because the quantities and qualities of body loading are what can create a body that is well, we need a grid—a way to easily measure the body and what it can do. Through a better understanding of alignment, we can break down and reconstruct our relationship with movement.

Alignment uses a basic grid or coordinate system in conjunction with the *anatomical reference system.* The anatomical reference system is a language created by anatomists and used by just about everyone who examines or treats the body to communicate general position and motion. There are decades' worth of data collected on the human body, both from cadaver and *in vivo* studies, that give us a good sense of what modern and historical bodies are and were capable of doing, load-wise. Placing a static or moving body along this grid, we are able to collect data based on the geometry of the body.

USING THE CORRECTIVES

Teaching exercise seems like it would be easy, right? Nope. Movement is incredibly complex, and as I've already mentioned, the details you can consider when prescribing exercise are endless. Even if we have never met, you and I are about to enter a teacher-student relationship, only I can't give you feedback on how you are performing the exercises that follow. I'll try here to be as specific as I can, but I'm going to ask you to assume full responsibility for listening to your body.

The following section is about the correctives in general—how they're set up, how the exercises can be changed to meet your body where it is now, and how you can turn this finite list into a set of infinite loads.

• Alignment Markers

These corrective exercises are to be done within the specific alignment parameters given, so that a grid can be used to determine where the loads are being placed.

• Rainbow Loads

Using a "rainbow" load means that once you've mastered the more linear version of an exercise, you can perform a similar move over a range of joint angles. For example, instead of pointing your foot directly ahead as in the first version of the Calf Stretch, you can turn your foot out three or four different angles and inward three or four different angles, stopping to stretch at each of these positions.

A rainbow of joint angles assists in creating the ranges of loads you'd experience if you were interacting with nature. When you begin rainbow loading, remember to drop back on the intensity and frequency of your correctives. Even if you've been doing the regular version of an exercise for some time, every new joint position will create a new load, never experienced before.

• Au Naturel

You can do these correctives in the comfort of your own home, but I will also include a more natural version of some correctives to help you see where they fit within the context of natural movement. This way you can slowly incorporate these movements into your regular life, and over time your need for a structured program can fade.

Keep in mind, the natural version I show you will be *the most advanced version.* You might find that years after reading this book, you are still making progress doing the more contained versions of this exercise. In some cases, you might have certain ailments or strength patterns that contraindicate the natural version. Don't let this bother you. Sharing the sentiment of Mark Sisson, paleo nutrition pioneer, it's about progression, not perfection.

• Progression

All the exercises are arranged in order from low to high loads. Start with the low loads and, over time, work toward the high. If you feel a stretch or exercise is intense as is, there is no need to progress further. There is no best frequency or duration of these exercises. Greater frequency (spreading the repetition throughout the day rather than doing it twenty times all at once) tends to yield better results. Unless, of course, you're doing these moves excessively and intensely, and not heeding the physiological signals of pain.

· Timing

With exercise, we are used to asking how long to do something, and how many times, but when it comes to these correctives, there is no pat answer to those questions. Hold a stretch as long as you feel like it. Do as many as you feel you need. The end goal of this book is to undo habits of being stationary and repetitive tasking, and there is no data on the minimums necessary to build new pathways.

Instead of approaching the upcoming movement sections as an exercise program, think of them as a new language you are learning to speak. After studying for five years, I was a kind-of-okay Spanish speaker, but after living for a month in the Yucatan, I was dreaming in Spanish. My brain took over and I didn't have to try so hard. Approach the rest of this book with the interest in becoming *fluent* in movement. Physical posturing is one of our most primal ways of communicating, and it's time to learn to "say" what you want to via the shape of your body!

YOUR FEET, SITTING, AND STANDING
CHAPTER 6

All the information in this book is essential. But if you only follow the protocol of a single chapter, let it be this one. No matter what your current physical status is, you can benefit by adjusting three things: your footwear, how you sit, and how you stand.

HEALING YOUR SOLE

I've hashed out the chronic immobilization sitting brings about, but even a ninja sitter has to get up to go to the bathroom or get something to eat every once in a while. Your feet almost never get a break from the immobilizing shoes they're shoved into. All day long, your toes are pressed together by the walls of the shoe. All day long, your ankle is slightly plantarflexed (with toes pointed) by the presence of a heel. (Any heel, even those sensible half-inch heels, the tiny rise on the heel of your "flats." And in case you're thinking that you don't wear heels, why don't you take a measuring tape to your favorite pair of tennis shoes?) Years spent gripping to keep a flip-flop or slide-on shoe attached to your foot create bent, fused toe joints. Since just about birth, your feet have had little opportunity to work their muscles, and as a result your lower-leg muscular strength, bone density, and nerve health have suffered.

But it's not only the fault of the shoe. As I wrote in the preceding chapter, the constant presence of a flat and uniform surface creates an immobilization effect by default. Walking repeatedly over even ground prevents all other joint movements besides the ones necessary for flat-ground walking. Every other joint configuration that your foot and ankle are capable of has become sticky.

Feet are extremely dexterous, not just to hold you up and help you navigate terrain; the sole of your foot is, like your nose or eyes, a sensory organ. The bony distortion created by stepping on something creates a neurological "image" in the system of your body that maintains an awareness of your position. Joint distortion, just like hearing, seeing, and smelling, gives you input about your environment for the purpose of eliciting a response from your body. The extreme number of sticky spots in modern human feet interferes with the communication between your body and brain, and in the case of walking or standing, your body's postural adjustment system can communicate inaccurate information about the environment.

For example, imagine someone is about to step on this pile of stones.

A supersupple foot is easily deformed by the stones, creating a unique shape. Information about this shape will be relayed to the brain by the sensory nervous system, where the brain generates an image of "what I'm stepping on."

The clear transmission of input to the brain allows the brain to integrate that information and respond with the most subtle shift of the ankle, knee, hip, or pelvis—a shift that allows the rest of the body to move forward without significant distortion or acceleration (both precursors for falling).

Now let's imagine that someone without much mobility in their foot steps on the same pile of stones.

The image created in this person's brain does not match the environment well.

The image is distorted, as is the reaction of this person's body to the surface of the stones. For many a sticky-foot, small obstacles like a slight crack in the sidewalk or a small dip in the floor can induce a fall or a sprain in the ankle because the foot has lost its sensitivity and mobility, and thus its ability to tell the brain how to direct the rest of the body in response to the information it's gleaned. When sticky spots in the foot are coupled with sticky spots in the hips, knees, back, and shoulders, moving can be extremely painful and resemble a series of lurches rather than a series of steps.

IMMOBILIZED BY SHOES

Years of sentencing their feet to shoes leaves people with significant atrophies in the muscles of their toes and the muscles between the bones of the foot (which are the arch-shapers), as well as a (semi-) permanent shortening of the Achilles' tendon and calf muscle group.

The interaction between the foot and footwear is so complex, I've written an entire book on it. Here, I'm going to skip the lengthy details of what's wrong with our feet and shoes, and focus instead on foot movement, specifically the correctives needed to undo the effects of immobilizing the toes, the individual bones of the foot, and, most importantly, the backs of the lower legs.

- **Calf Stretch**
  Foot marker: Use these approximate points to line up your foot forward.

Place a thick folded and rolled towel on the floor in front of you. Step onto the towel with a bare foot, placing the ball of the foot on the top of the towel and keeping your heel on the floor. Adjust the foot so that it points straight forward. Keeping your body upright (shoulders and hips over heels), step forward with the opposite foot.

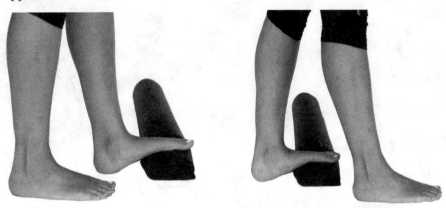

The tighter the lower leg muscles and tendons, the harder it will be to bring the other foot in front. Many have to keep the non-stretching foot behind the towel. Many can bring the opposite foot forward but only with lots of butt, quad, and jaw clenching. Only step as far forward as you can while keeping everything else relaxed. Repeat other side.

For a greater load, use a phone book or half foam roller (the one pictured here is three inches in height).

The higher the object you step on, the greater the load and the more challenging it will be to get off of the non-stretching leg. Progress wisely.

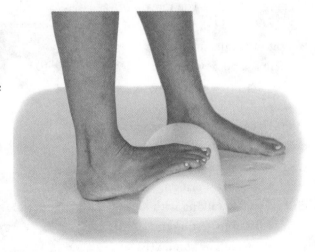

- **Rainbow Calf Stretch**

Starting with the foot turned out 45°, step up onto a rolled towel (or dome) for a Calf Stretch. Turning in at 5–10° increments, load this joint angle via the basic Calf Stretch parameters.

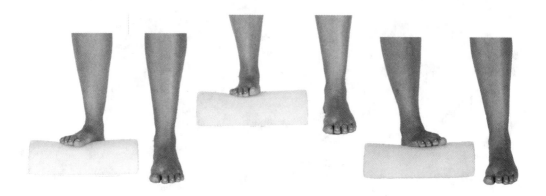

- **Au Naturel**

The loaded dorsiflexion (smaller than a 90° ankle angle) created in the Calf Stretch is naturally found walking uphill and stepping on debris as you negotiate natural terrain. While hiking uphill, pay special attention to your back heel, keeping it down as you move forward (instead of keeping your weight forward, on your toes).

There is a point when the heel comes up naturally, but due to years of shoe-wearing, this moment comes faster in the gait cycle than it should. You can also Calf Stretch on various logs or rocks when meandering, especially if you're walking with kids.

In addition to getting this stretch outside, you can do the Calf Stretch in your home on

whatever is lying on the ground. By utilizing the wildly varying terrain that is a toddler-filled home, for example, I am personally able to find unique stimulations throughout the day—no nature required. I do not recommend Calf Stretching on jacks or Lego. Especially in the middle of the night. Unless you don't mind letting out a bunch of curse words.

· Top of the Foot Stretch

Years of tight calves mean years of excessive tension between the shin and the foot. This stretch applies a tensile load to the tissues between the foot and shin as well as to the muscles between the toes and foot. Note: If you have hypermobile ankles (when you point your toes, your ankle joint reaches 180° or more), focus the stretch between the toes and foot, not between the ankle and shin.

Stand up, barefoot, and reach one leg behind you, tucking the toes under as shown, making sure to keep the torso upright (it is common to move the pelvis or upper body forward).

Tight "gripping" muscles can make the foot cramp during this exercise; to avoid this, try doing it more often, but for a shorter time. Also, to reduce the load, try doing this exercise while sitting or reach the leg behind you less. For a greater load, step farther back with your stretching foot.

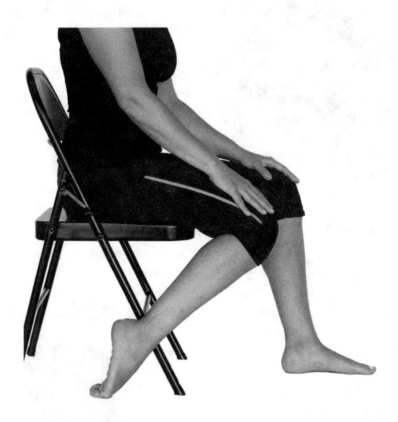

- **Rainbow Top of the Foot Stretch**

  You can rainbow this stretch in two ways. The first is to present a different aspect of the ankle by rotating the thigh to varying degrees, as pictured. The second is to allow the ankle to collapse to the right and left to varying degrees, as pictured (opposite page).

- **Au Naturel**

  The Top of the Foot Stretch is helpful in making sitting with the legs folded under possible, a resting posture common in many chair-free populations. For more examples of positions, see page 106.

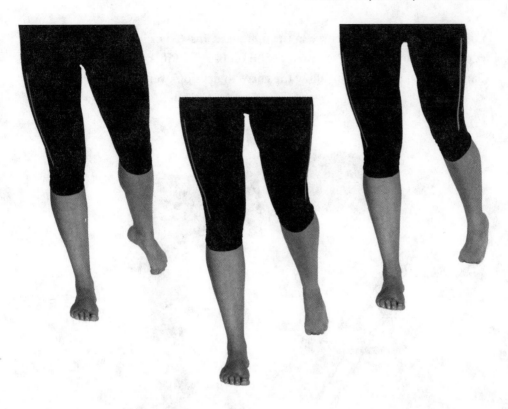

- **Double Calf Stretch**

A constant, slight knee flexion is a common adaptation to positive-heeled foot-wear. And even if you don't wear a lot of heeled shoes, "keep your knees slightly bent" is a commonly cited postural guideline. While perhaps accurate in certain situations (like standing, for an hour, holding your tuba, in a marching band), chronic knee flexion is not optimal for the human knee.

But knee flexion doesn't only shorten the calf muscles, it shortens the muscles on the back of the thigh as well. Over time, this shortness can pull the pelvis into a posterior tilt, which also tends to flex the lower vertebrae in the spine. The Double Calf Stretch intensifies the loads found in the single Calf Stretch and also involves the hamstrings, especially if you untuck your pelvis, creating a small curve in the lower back.

As with all the exercises, forcing a small curve in your back is not ideal—force-fully extending the spine to create a curve is not the same thing as releasing the muscles that removed the curve in the first place. Instead of lifting the tailbone, let gravity drop the ASIS (see anatomy bit on page 113 if you don't know what that is) toward the floor and allow the curve to develop over time.

• **Foot Bone Mobilization on Ball**

   Mobilizing the bones between the feet is a bit more challenging than the other exercises listed here. Because the numerous muscles running between the foot bones are not big movers (like your arm and leg bones are), they are mobilized better through stepping on lumps and bumps. But stepping on lumps and bumps when your feet are stiff can be very painful. So to keep excessive loading at bay, start by draping your foot over a tennis ball. Move your foot a bit forward. A bit backward. A little to the left, and a little to the right. Use the ball in the same way you use the vacuum—systematically reaching every inch using a grid-like progression.

Doing this sitting yields the smallest loads. Standing, greater loads. Standing and pushing your foot down onto the ball, the highest load. Other ways to play with the loads are to change the shape and density of the ball. The smaller and firmer the ball, the higher the load. (My favorite balls to use are those designed specifically for this purpose. You can find my recommendations in the Appendix.)

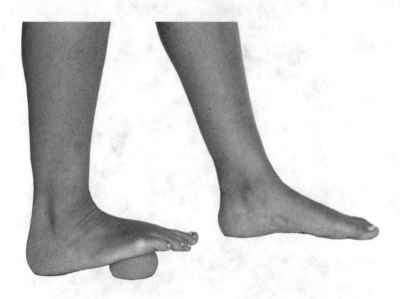

• Rainbow Foot Bone Mobilization

Cobblestone mats are a great way to "walk" on rocks in the comfort of your own home and make a great addition to a standing workstation. You can also make a cobblestone path, inside or out, by making a small frame and filling it with sand, stones, and rocks of various sizes. Walking on this a few times daily provides stimulation to foot musculature that in turn adapts by becoming stronger and better able to handle these forces for longer periods of time.

- ## Au Naturel

To generate these loads naturally you can, of course, walk outside in your bare feet. More "barefoot" spaces are popping up all the time—spaces where dangerous debris is minimized, terrains vary, and there is a large space for foot stimulation. In addition to the grounding benefits of foot-on-Earth interaction, natural terrain also offers the best place for skin adaptation.

As you will read in the hanging chapter, it is often our skin weakness that prevents us from recruiting our muscles in a natural, synergistic fashion. Toughen up your feet by walking on terrain that yields to your foot in different ways. The callus that will develop from walking barefoot outside will have the best circulation and the most cellular activity of any part of the foot. Corns and calluses are only problematic when they reside right next to unadapted skin. The way an isolated callus feels may be problematic at first, but the callus is your ticket to more robust movement. Work to develop calluses more evenly and use them as a biofeedback tool. When I do a lot of hanging, my index fingers develop fewer calluses when compared to my middle finger. This lets me know that I don't grasp as much with my index finger, so when I hang, I'm mindful about working that finger more so as to distribute the work to my upper body more evenly.

A funny story: When my husband was in Thailand, he got a Thai foot massage and pedicure. Blissed out and half asleep through most of it, he woke at the end in horror, realizing that his masseuse had cut off all of his calluses—calluses he had built over decades. He was unable to walk "normally" for about a year, until they built back up again. His entire kinetic chain—the muscles and the patterns in which he used them—depended on the interaction between his foot (the foot he had *before* the session) and the ground.

I predict that, in the future, movement scientists will see the errors in assessing the kinetic chain without considering the interaction between the skin and the ground.

GROUNDING

Footwear impacts more than just your feet! Newer data shows that footwear not only alters the function of the knees, hips, and spine, but also can impact the process of conducting electrons between the ground and body. *Grounding* (also known as *earthing*) is the interaction between the body and the Earth's surface electrons. Research shows that earthing can affect physiological outcomes— like better sleep and decreased pain. Barefoot time, in fact, might even be *essential* for us humans to function optimally.

• Feet in Daily Life

There are things *to do* that make your feet more mobile, like the exercises listed above, and then there are things that can be done without much effort. The first is to pick better footwear. You can spend twenty minutes stretching your calves every day, but if you keep putting shoes on that shorten them back up again, the ratio of time spent lengthening to shortening your muscles is not in your favor. By picking a heel-free shoe, you are essentially loading these tissues more optimally than if you were wearing a heel. The same goes for toe-spreading space.

If your current shoes limit your toes' range of motion, then you are practicing toe-binding more often than not. Get wider shoes or those flexible enough for you to spread your toes while sitting, standing, or walking. Foot-alignment socks are another way to get your stretch on while you do other things. Slip these puppies on while watching TV, making dinner, or sleeping, and you can up the amount time you are "working on your body" without needing to schedule additional time.

SITTING

You are going to sit a lot of the day. Which is fine. It's entirely natural to sit and rest and hang out with your peeps. But there isn't any requirement that you spend your sitting time in a chair. It takes no additional time to sit on the floor instead of on your couch. What do you gain from sitting on the floor? First there are the numerous ways you can position your joints—each one creating a unique load.

Anthropologist Gordon Hewes was one of the first (and unfortunately one of the few) scientists interested in human body postures, communication, and culture. As a result of his interest, he spent years compiling how people take their rest time all over the world. Here is an adaptation of this research:

But sitting on the floor isn't as easy as some people think! Not only have we lost the mobilities created by sitting, our bony structure has also adapted to chair-sitting, making some positions unavailable to us without a major overhaul in habit.

Even something as simple as "sitting on the floor" can require assistance. To help mobilize your hips and knees for sitting on something other than a chair, try the following stretches. Also, it is helpful to meet your body's ranges of motion where they are right now. This means floor pillows or rolled blankets under the hips will be a staple for most!

• Sole-to-Sole Sit

Sitting on a pillow or folded blanket (choose a thickness that makes sitting this way comfortable), place the soles of your feet together, letting your knees drop away from each other. If your pelvis is forced to tuck under, increase the height of the blanket until your pelvis can tilt forward (like your pelvis is a bowl of soup that you're dumping in front of you).

• **Rainbow Sole-to-Sole Sit**

You can lean forward to stretch your groin a bit more, as well as turn your torso to the right and left to change the load profile. Great for playing games with your kids.

• **Cross-Legged Sit**

From the Sole-to-Sole Sit, cross one shin over the other. Try different combinations of leaning forward and rotating, as well as changing the distance between your feet and groin. Slowly "paint" an imaginary circle around you on the floor. Twist, lean forward, and do your best to keep both sides of your bottom anchored to the floor. Repeat, crossing the other leg in front.

• **Soles Against a Wall**

Place the soles of both feet against a wall, making sure to elevate your hips until you can straighten your knees comfortably. Relax your body toward your thighs, no forcing, no bouncing. Want to feel something really amazing? After relaxing your torso as far as it can go, relax your head toward your thighs. By decreasing the tension down the back of your neck, you'll increase the tension all the way down the spine. This stretch goes beyond individual muscles and is a great way to load the connective tissue wrapping that surrounds the muscles of the legs, spine, and head!

ALTERNATE-TO-SITTING WORKSTATIONS

There's a big movement afoot to start using standing workstations, and I definitely love that everyone is waking up to the dangers of sitting all day. But I want to stress that we can't just swap out standing for sitting and assume we've solved all our body problems.

If you DO decide to stand for most of your workday, keep a few things in mind. One is that you don't need to pay a lot of money for a fancy tall desk—standing workstations come in a range of prices and materials, but they are a pretty basic concept. You can take your computer to a high counter, flip a box upside down on your ordinary desk, or work at a bar table, right now, without having to save up or wait for a desk to ship. Also keep a rolled towel or a half foam roller by your workstation, so you can take advantage of your standing time to do Calf Stretches (see pages 96–99) and Pelvic Lists (see chapter 9). Shift your weight often. Consider getting a cobblestone mat to stand on, to keep your intrinsic foot muscles working while you are. And, of course, take frequent movement breaks.

But remember that *any* workstation that is not the standard office chair is beneficial, and the more you change your position, the better. If you have a laptop, you can work in so many different positions—sit on the floor with your computer up on a box, or lie down, or kneel with your pelvis untucked. Change your position frequently. Most of all, as always, move—as much as you can.

- **V-Sit**

This stretch is right out of elementary PE class, but it's effective! Sitting on a small pillow, widen your legs until you find your groin's limit and hang out there for a bit. Lean forward as you paint a "rainbow" shape from leg to leg. You can take your time in each spot as well as move smoothly from side to side. Each of these (static holds and stretching while you move) creates unique loads.

- **Sitting in Daily Life**

We are so in the "exercise" mentality that we tend to associate unique body positions with a specialized setting—like we need a class, an instructor, a sticky mat, and a special hour to move and sit in unique ways. If you already take an hour to do this, great, but don't undo your work by practicing your most-practiced habit, sitting on furniture. Instead, sit on the floor for meals and while reading, chatting, or just watching TV. No additional effort needed. Well, maybe a little; getting down to and up from the floor uses a greater range of motion of your legs and in turn keeps them strong enough to do so. You won't know you've lost the ability to get down and up from the floor until you try it one day and surprise! It's not as easy as it once was. An interesting note: The more you need to use your hands and knees to get up from the floor, the greater your risk of dying from all causes. Want to stay functionally strong? Get up and down daily and sneak a nip of exercise into the living of your life.

STANDING

You probably spend a lot of time standing around. At the grocery store. At the bank. While chatting with people at work. And, if you have kids, perhaps along the practice or game sidelines. Standing, more than sitting, loads your body with its weight. But *how* you stand dictates where that weight is, which, in turn, dictates which tissues are taking the brunt of the load.

There are certain loading patterns, or postures, that burden the body by affecting how the muscles can participate in distributing the load. Below are variations on a common stance, brought about by excessive sitting, positive-heeled shoes, and cultural reinforcement.

Note, in this series of pictures, the different ways a pelvis can twist and jut out in front, the ribs and chest lift, and the feet turn out. Again, like sitting, there isn't really anything wrong with this particular position; it's just that you've used this single position most of your life and your tissues are suffering for it. Begin by making these adjustments:

• Straighten Your Feet

Line up the feet so they look more like the tires on your car when you're driving forward. When we get to gait details, I'll be even more specific, but for now, try to correct some of your turnout. Years of walking a certain way (flat ground, positive-heeled shoes) have changed the soft tissues of your lower leg so this adjustment comes, initially, from turning the thigh bones in. Which means once you've adjusted your feet, you'll have to correct the thighs. So...

• Externally Rotate Your Thigh Bones

The first time you do this, it's best to use a mirror. Roll up your pants to above your knees. With the mirror behind you, bend forward and look at the backs of your knees. I like to call these your "knee pits." With your feet pointing forward, your pits are likely on the sides of your legs. You need to get them back where they go—directly behind you. To do this, you'll have to rotate the thighs external-ly—a motion that sends your kneecaps away from the midline and your knee pits toward it.

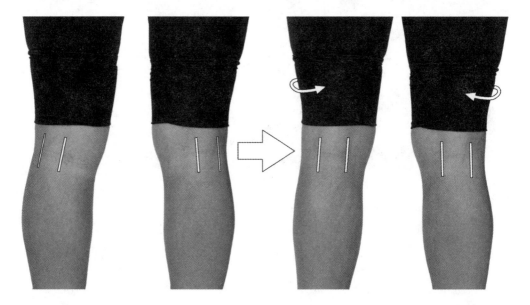

Ideally you'll want to use your deep hip rotators to create this motion, although initially you'll probably tense the quads to do so. So...

ANATOMY BIT

The pelvis is made up of three bones—two iliac bones (one on the right and one on the left side), and a sacrum in the back. The anterior superior iliac spines (ASIS) are the most prominent anterior (front) superior (above) bony projections on the right and left side of your pelvis. People often refer to these points as the hip bones (as in, "Put your hands on your hips").

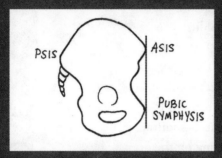

The pubic symphysis is the joint at which the two hip bones come together. It is the lowest bony prominence before your pelvis wraps around to the undercarriage.

• Find Your Neutral Pelvis

When you're standing, a forward pelvic thrust creates more work for the front of the thigh (quads, iliacus, and psoas) than it does for the muscles of your backside. Share the work required for standing more evenly by backing up your pelvis so that your hips align vertically with your knees and ankles. (Image next page.)

Once you've backed your hips up, you can find your pelvis's neutral position. Some people define this as aligning the posterior superior iliac spine (PSIS) and the anterior superior iliac spine (ASIS) horizontally, and others arrange the ASIS and the pubic symphysis vertically. While there is a slight difference between the two positions, I prefer the latter. The markers on the front of the pelvis are easier to feel and see—much more user-friendly—while the posterior markers are often located under thicker layers of soft tissue, not to mention behind you. Using the ASIS and pubic symphysis makes it easier for you to correct yourself throughout the day. And now you are in the perfect position to…

· Relax your kneecaps

Once your hips are back, the front of your thighs will be used less and your quads can relax, dropping your kneecaps toward the floor—we call this the Kneecap Release. This isn't as easy as it seems! Your quads might have been involved in holding you up the last few decades, so give them time to drop. To take all the work away from your quads, rest your hips back against a wall and "turn off" your thighs for a moment. Sometimes leaning forward while being on the wall helps too.

Once they're down, see if you can stand upright without them switching on again. Because time is needed to develop strength in your backside, you might find yourself feeling like you're falling back when you're not thrusting forward. Small loads (read: backing up a little bit at a time, over a month) should give these muscles time to catch up!

· Standing in Daily Life

Each of these adjustments undoes sticky spots common to excessive screen time, and as you follow the other exercise and movement guidelines in the rest of the book, these corrections will become easier and more second nature. You won't be

creating the tensions that push you out of alignment and you'll be developing a more supportive muscular tone. Making these simple adjustments over and over again—while you are just standing around—can improve the loads to the lower back, hips, knees, and pelvis *in an instant*. It's not like you were doing anything anyway. You were just standing there, right?

MOUSE HANDS TO MONKEY ARMS
CHAPTER 7

The beauty of the human body, in my opinion, is how using it can result in tissue adaptations that make it even more usable. For example, muscles contracting and relaxing to climb up a tree for food aren't only facilitating the input of calories. The pull of working muscles transfers to their tendons and bony attachments, which in turn causes an increase in tendon and bone mass and strength. This increase in mass now requires you to eat more calories, but via this process you've become more physically capable to gather food. The relationship between nature and our physiology continuously regulates itself in a natural context.

The use of the arms is particularly important both to making our body more usable and to the maintenance of the body's structure. Upper-body tone not only keeps the joints of the shoulders, elbows, and wrists stable and operating smoothly, optimal tension in this area is responsible for keeping the upper (thoracic) spine upright.

YOU'RE USING YOUR ARMS TOO LITTLE! AND TOO MUCH!

In addition to being chronically underused, our arms tend to have deep-set asymmetries all the way to the bone as a result of our right- or left-handedness. In a more natural environment, not only would you have used your hands, arms, and shoulders way more, you would have also used a *single* arm way less often.

As in my earlier example of kwashiorkor (a disease that arises when the ratio of nutrients is off), our ratio of arm use is off. Yes, the total amount of arm use is low in both arms, but the habit of using a single arm is also high. Think about all the things you do with your arms each day. Brushing your teeth. Using silverware. Jotting down a note or two. Opening and closing doors. Using a computer mouse. In every one of these instances, you're likely to use your dominant hand. And, to keep that dominant hand free, you're likely to carry your gear—your purse, your shoulder bag—on top of the opposite side's shoulder almost all of the time. Handedness and bag-wearing isn't a problem now and then; it's the length of time you've done it and the lack of doing everything else that lead to tissue alteration. In the case of your upper body, you've developed asymmetrical tissue strengths due to repetitive, low-force patterning in one arm, and you've got significant weakness in both.

Have you ever seen an older, grandmotherly woman shuffling along with her spine curved forward into a dowager's hump? This forward curl of the upper spine is called *hyperkyphosis,* and many associate this posture with age or a deterioration of the bones of the spine. But as I will illustrate here, *most* people are walking around with a hyperkyphotic spine, and have been for decades.

Before I go on, think back to our earlier example of the orca whale. The structural integrity and function of the orca's fin was maintained *indirectly* via moving a certain way through a certain environment. The collapse was not due to weakness of the fin muscle. There *is* no fin muscle, and in his natural environment, an orca does not require a specialized fin-holding muscle. In fact, a specialized fin-holding muscle would be a waste, as the fin's structure is maintained by the forces created by other muscles moving in a certain way through a certain environment.

Just as natural whale movement keeps fins up, the strength created via natural movement keeps the spine up. In nature, you would have to use your arms, all day, to: forage (digging, picking); hunt (making and throwing weapons, carrying carcasses); process food (oftentimes beating items literally to a pulp); build structures; carry wood; carry water; carry your kids; carry your worldly posses- sions; clean animal hides; clamber (for fun, safety, or travel).

Wildly varying muscle-use patterns created by ever-changing scenarios are much better at engaging the upper body's full range of motion, which should cover a nearly spherical field.

To see what *your* upper body is currently capable of, pretend your upper body is surrounded by a globe. Reach your arms up. That's the top of the globe. Reach down. There's the bottom. Reach your hands out to "touch" the sides of the globe. Now use your fingertips to "paint" as much of the globe's inner surface as you can. The more you do this, the easier it will become and the more of the globe you can touch. (Yes, there is a big part of the globe behind you that's not "paintable," but I'll bet you can improve what *is* possible to paint by practicing this as well as all the other stuff in this chapter.)

Once you've met your shoulder girdle's field of motion, consider your most frequently used shoulder motion or position. For most of us, this isn't spher- ical at all, as our hands spend most of their time fairly stationary out in front of us—as found in school classrooms and keyboards the world over.

In addition to how much of that globe you use a day, consider how much of your upper body's potential you have used over a lifetime. It's not a physiolog- ical mystery why or how our soft tissue has frozen parts of the neck, arm, and shoulder joints out in front of us. We literally live our lives there, and thus have adapted to this narrow and frequent use pattern.

Over time, this added weight out in front of the body increases the drag on the spine, causing it to spill over. Which is why parents, teachers, and other adults around the world instruct children regularly to stand up straight—a verbal cue that typically results in the retraction of the shoulder blades, and a lift of the chest and chin.

While these adjustments (chest up, shoulders back) reduce the forward-dis- placing loads to the spine, they don't actually undo the curve; they just hide it. And, even worse, adjustments made to facilitate a temporary visual improve- ment actually introduce new curves in the opposite direction and compromise the mechanical leverage of the muscles that support the spine—all leading to an *even greater* curve over time.

KYPHOSIS IS OUR FLOPPY FIN

A forward-falling spine is similar to the captive orca's fin in that it is a collapse allowed by the absence of certain kinds of movement. The spine, like the fin, does not hold itself up directly, but rather calls on the surrounding muscular tones and tensions—a particular pattern of strengths that arises from moving in a natural way.

Our arms are attached to our shoulders, which attach to our scapulae (shoulder blades), which attach to the spine—but you could just as easily say that our spine attaches to our scapulae to our shoulders to our arms to our hands. The musculature in the arms should be strong enough not only to bring things toward the body (using your biceps to pull a weight toward you), but also to bring the body closer to the hands, as it does when you do a chin-up.

Greater use of the upper body—and by "greater" I mean pushing, pulling, lifting in all different ranges of motion—creates constant, low-grade tensions between the shoulder blades that stabilize the spine while moving, and support it (read:

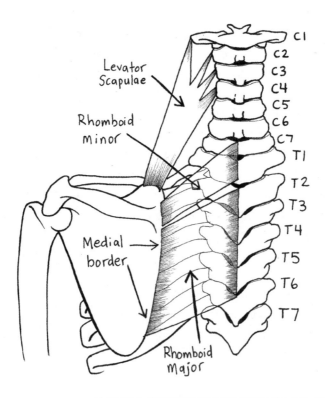

prevent it from collapsing forward) at rest. The indirect result of this use is a constant resistance to the forward collapse of the spine. But in order to keep the spine-supporting muscles between your shoulder blades (rhomboids) generating force, *you have to stop pulling them together all of the time.*

Using a bird's-eye view, not-to-scale model of the shoulder girdle, anatomical neutral looks something like this.

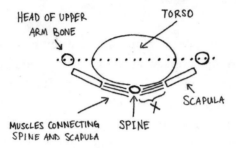

The medial (closest to the spine) edges of the shoulder blades are rarely too far away from your spine. The problem in "forward shoulders" is that the lateral edges (the tips of your scapular wings) are too far forward and the humeral head (the top of the upper arm bone) has rolled inward and off of the scapula.

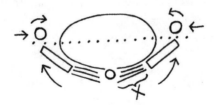

A lot of the time, one side is more forward and internally rotated than the other.

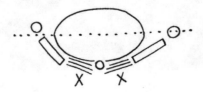

Which means when you pull your shoulders back to make them even, you've got muscle tensions that differ from right to left.

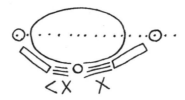

By correcting your posture, you have introduced a chronic position that can change the muscle mass and leverage on one side compared to the other, resulting in loads that can create a "permanent" twist in the vertebral column over time.

Were we moving our arms as nature requires—throughout a day, throughout a lifetime—we wouldn't need to "correct" our posture; the segments of our body would be maintained by this use. Which means that what we do now to correct our forward slump *isn't really a correction at all*. And while we can buy a little time by drawing our shoulder blades back, the spine and the tissues of the shoulders and arms cannot function for long if we don't actually fix the issue.

The first thing you can do to improve the strength of your upper body *and* the stability of your spinal column is to stop hiding your kyphosis. Add these to your lower-body alignment checkpoints when standing around:

- **Ribs Down**

Drop your rib cage so that the lowest, most forward bony protrusion of your ribs is stacked vertically over the highest bony protrusions on the front of the pelvis.

ANATOMY BIT

The highest bony protrusions on the front of your pelvis are clinically referred to as your anterior superior iliac spine, or ASIS.

Bringing your ribs home also adjusts the lengths and optimizes the leverages of every muscle running between your pelvis and rib cage. This simple adjustment can improve the state of your abdominal muscle groups and lower back, and sets a stronger foundation for your upper body.

• Shoulder Blades Neutral

If you reach your arms forward like you are hugging a tree, you can probably find the medial border of your shoulder blades. This border doesn't need to be superfar from or superclose to the spine, but should reside somewhere in between.

Start by squeezing your scapula all the way together. Using one hand over the opposite shoulder to feel behind you, find your shoulder blade. Slowly allow your scapula to relax and widen just to the point where the medial border is no longer poking out. There should be no bony protrusion—the medial border is now in the same plane as your spine. In most cases this just takes relaxation, but in some—especially in those who have extreme chest tension—you might have to actively pull your shoulder forward to smooth out this edge.

Now, with your ribs down and the medial border of your scapula flush, take a look at your spine in a mirror. This is the curve of your thoracic spine. This curve is not altered by bringing your shoulder blades together, nor is it altered by lifting your chest. This curve *is always shaped as it is right now*. And it probably needs to be reduced, which is what we will work on for the remainder of this chapter.

WAIT. AM I SUPPOSED TO WALK AROUND LIKE THIS?

Once you have your rib cage and scapula in line with the rest of your body, you're probably looking (and feeling) a little droopy. Which always leads to the question, "Am I supposed to keep my body in this position all the time?" The answer is no, of course not. But at some point you have to undo the masking of habitual positioning if you are going to change it. You can't get your muscles to work a new way, to support the thoracic body, if they are always occupied with hiding your poor form. So the process is more like this: Relax some of your mask (drop your ribs and relax your shoulders when you remember). Do a little corrective work to change the mechanics of the shoulder. Relax the mask some more. Do a little more work. Slowly toggling between these two will gradually reshape your body.

ANATOMY BIT

Shoulder Girdle Smoke and Mirrors

In trying to reduce the appearance of this forward shoulder-slump, we move the *scapulae* in order to bring the shoulders in line with the ears. But we didn't change the mechanics of the shoulder joints themselves in this case; we've simply moved the wonky joints relative to the ground. The interaction between the parts that make up the shoulder joints is the same as before.

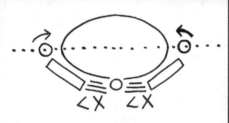

The gross curve of your spine is created by the position of the vertebrae relative to each other. In the case of hyperkyphosis, the vertebrae can be rotated or slightly sheared forward, which means that these are the motions necessary to UNkyphosis. (Image bottom left.) Lifting the chest is simply displacing the entire curve relative to the ground.

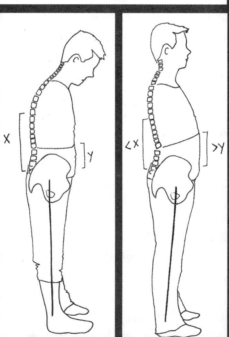

The curve is no longer visible, but it still exists—and now the entire rib cage (and all its function) has been rotated, creating unnecessary forces and decreasing necessary ones essential to the thoracic cavity (home to your heart, lungs, diaphragm, and spine).

GOING FROM MOUSE HANDS TO MONKEY ARMS

Question: Is upper-body strength a big deal?

Answer: No. Upper-body strength is a *huge* deal.

This section is going to help you transition your computer-savvy arms to arms that can move you (and other stuff) through life without hurting. But in the end, the level of function we are after here goes beyond lifting heavy things. In fact, it is the shoulder girdle's role in moving *tiny* things—molecules of oxygen—that we will eventually affect most. Let's begin with the weakest part of your upper body: your hand skin.

There are a lot of visual signs that give you insights about how you use your body. Calluses (or a lack of them) offer up information about how you've been interacting with your environment. Take a look at your hands and read the story your body is telling. Do you see evidence of your lifelong habits?

A callus is an area of the skin receiving better blood flow compared to the rest of the hand or foot. A callus is problematic when this well-vascularized spot sits right next to an underused area. Repetitive patterns (like always having the same grip on a kettlebell) can make one spot on your hand much stronger than the rest of the hand, which is then too weak to hang on to the strong bit. In this case a callus will eventually rip off (ouch) if you continue to load your hands in exactly the same way. A small callus on your foot is like stepping on the same small rock over and over again. This repetitive pressure in the same location deepens the callus in just that one place to the point of literally creating a "stone-like" object inside your foot that can destroy neighboring tissue and become an ulcer. Now the callus *is* a problem. (But again, the callus isn't really the problem—it's the lack of callus everywhere else that is! See how so much of our medical diagnosis is based on perspective? Should a medical professional look at your deep callus and say, "Whoa, this pain you are having in your foot is because you haven't been using your foot skin nearly enough," or should they say, "Whoa, you have a case of hyperkeratosis and I can cut that out, no problem"? Neither of these statements is wrong, per se; but one gives a more clear representation of the problem. I digress.)

Believe it or not, when it comes to closing the upper-body strength gap, what gets in most people's way is the weakness of their skin. Skin does not have muscle, but it actively thickens over time in response to different types of loading. Skin can adapt to pressure, but what really makes it thicken into a callus is a shear force.

- **Toughen up**

To prep your skin when you decide to pursue upper-body strength in earnest, start with small loads, pretend-hanging (this is where you keep your feet on the ground but bend your knees enough to feel the weight of your body press your skin into the surface of whatever you're hanging on). Each surface creates a different load, so bars of various textures, branches, and even the doorjambs in your house can all contribute to the skin strength you will gain. Even before you start hanging, you can work skin-training in here and there. Eventually your hand skin will be a full-fledged member of your kinetic chain—which will come in handy (get it?) later.

GIVE ME FIVE

Giving fitness evaluations was part of my list of duties working in the exercise lab in college. Part of these evaluations included a hand dynamometer—a device used to measure grip strength. Most people I measured fell into a normal range, but one time this older woman came in and practically pegged the thing! Startled by her performance, I began collecting data on her exercise practice and was surprised when nothing really stood out. We were just about finished when I happened to spot her business card: Cake decorator. Ah! That made sense. This woman did not *exercise* her hands more than everyone else; she just *used* them more in her daily life.

For most of us, keyboarding and texting make up our hand use. We don't dig daily, and for hours, with sticks. We really don't use tools much, and when we do we reach for those that are battery-operated or of the most ergonomic design. And we rarely, if ever, hang from our hands.

Pretend you're placing your hands onto a keyboard. Do you see how they form little upside-down cups? The cup shape is created by the sum total of your finger joint's flexion (forward bending). And there is often a semi-permanent extension (or backwards bend) at the wrist.

Restore your wrist and finger range of motion with the following:

- **Reverse Prayer Hands**

Place the backs of your hands together, so that all five fingers touch. If your lower arms are tight, getting the thumbs to touch will be the trickiest. Then lower the wrists until they are at the same height as the elbows—without separating the fingers. If the fingers do separate, reverse directions (take your wrists away from the floor) until they meet, and hang out there.

• Finger Extension

With hands palm up, place your fingertips onto the floor (if you are kneeling) or table (if you are standing) or hold them with your opposite hand and gently press the heel of your palm forward, away from the body. (Image left on top of page 128.)

Try to extend each finger joint (in other words, do the opposite of forming a fist), making sure to stop if any finger joint buckles back into flexion. (Image right on top of page 128.)

Once you have extended your finger joints, slowly bend your elbows, until they point directly back behind you and not off to the sides. Your fingers, too, should be directed straight back; a twist in the wrist (as shown by your fingers veering to the left or right) is a sign of tension.

AN ENTIRELY DIFFERENT WAY OF USING THE SAME EXACT MUSCLES

Imagine you are holding something in your hand. In this case, the muscles just above an area are responsible for holding the weight just below. Your fingers are loaded with the weight of the object; your wrist is loaded with the weight of the fingers and the object; the lower arm carries the weight of the hand, fingers, and object; and so on, all the way up the shoulder. But if we reverse this so that your hand is holding on to a bar and your body dangles below, these same muscles now carry everything *below*. Those same muscles of the wrist—the ones formerly loaded with the combined weight of the object and hand—are now bearing the load of the entire weight of the body (minus the hand, of course).

It is likely your tissues reflect your lack of regularly holding your body weight on your arms, throughout your life. Which isn't to say that you couldn't go hang right now, but that if you did, the microtrauma and the way you execute the hang on the cellular level would be different than it would had you been hanging throughout your life.

A STEPWISE APPROACH TO MORE WHOLE-BODY STRENGTH

To progress towards greater strength, a stepwise approach comes in handy. The best way to achieve your end goal of being able to haul yourself up and over something—a six-foot wall, for example—is to call on your whole body to participate, not just the

WHAT DO YOUR BONES SAY ABOUT YOUR UPPER-BODY USE?

Anthropologists use data on bone robusticity to create a picture of how our ancestors used their bodies. Since our bodies adapt to our behaviors, the nuances of bone are great indicators of what people have done over a lifetime.

Bony nuances in Natufian hunter-gatherer males tend to show heavy right-handedness, and adaptations to muscular attachment sites used in overhand throwing. Other populations show one-sided bony adaptations believed to be created by the pushing (spear) or pulling (bow) of a weapon, the strenuous work of hide-scraping (women) and food preparation (women).

In case you and your modern body are feeling left out of this seemingly ancient phenomenon (sure, people used to adapt, but haven't we evolved past all that?), rest assured, we continue to gather evidence on populations that move with higher frequencies and greater loads and how their bones end up looking.

Computed tomography (CT scans) of professional baseball pitchers show that bones in their throwing arm form to the forces created by pitching the ball. A pitch requires exaggerated wind-up that places a particular torquing load to bone that, over time, remodels the bone in the same way pressing a hunk of clay can change its shape.

parts of you that are most able to participate due to your modern-life habits.

While certain diagnoses like "hypermobility syndrome" imply that a person is hypermobile, people are *never* head-to-toe-hypermobile except in the most extreme cases of connective tissue disease. I've had plenty of people with diagnoses like this come to see me, only to have them demonstrate excessive mobility (better termed *joint laxity*) in only a few areas. The rest of their body hardly moves at all.

This phenomenon of some parts moving excessively while other parts don't move at all is common to all of us with hardly used arms.

• Thoracic Extension and Head Ramping

To experience what I am talking about, try this: Arrange your sternum so that it is, more or less, vertical, and lower your chin towards your chest. Now bring your head back up—the first time by lifting your head only. Here, the uppermost vertebrae of your cervical spine in your neck rotated posteriorly (backwards) to bring your eyes to the horizon.

Now arrange your sternum and lower your chin to your chest again. This time, lift your head up by "lifting" only your chest. Your head should not move relative to the body, but by simultaneously thrusting and rotating your rib cage you can, eventually, get your eyes to the horizon, with your chin still mostly down.

You can probably guess that neither of these ways of getting your head up is ideal. But most of us maintain our head position by using a combination of these two motions. The result is hyperextension of the neck (for those of you who wear bifocals, you know how constantly lifting your chin can cause headaches and neck pain), constant vertebral shear in the spine, inappropriate forces to the vertebral column, and, most pertinent to this chapter, the failure to work the spinal muscles that reside between the neck and the lowest rib.

The muscles of the upper back are our best allies when it comes to undoing kyphosis—yet most people have no clue that they're bypassing those muscles regularly. So, are you ready to meet the muscles of the thoracic spine? Here we go.

Sitting or standing against a wall, bring your sternum vertical once more. Lower your chin to your chest. Without lifting the chin *or* moving the rib cage, bring your eyes up the horizon. Think of sliding your head back to the wall behind you, but don't lift your chin. The movement here will be small at first, but you should, maybe for the first time, feel work in the muscles that move the kyphotic part of your spine.

• The Rhomboid Pushup

This is my favorite exercise to reintroduce movement back into the "sticky" middle spine. (Image on opposite page.)

Start on your hands and knees, letting your knees and wrists fall directly under your hips and shoulders. Let your head, pelvis, and belly relax and lower to the floor.

Slowly allow the torso to move toward the floor, which will bring the shoulder blades together.

Note: This is different from *actively* squeezing your blades together. You're trying to work with gravity here and *allow* the movement—not force it.

Once you have reached the "bottom" of the exercise, move the entire spine up toward the ceiling. This motion will spread the shoulder blades apart.

Do not round the upper back or tuck the pelvis. Repeat a dozen times, focusing on maximizing the motion between the shoulder blades.

This Rhomboid Pushup is often mistaken for a cat-cow yoga pose, but in the Rhomboid Pushup there is not a change in the curve of the spine. You aren't rounding your back and then extending it. Instead, the spinal column maintains its original curve as it moves toward and then away from the floor.

• Quadruped Hand Stretch

The way you position your hands when on your hands and knees reveals a lot about the tension in your fingers, hands, wrists, forearms, and shoulders. On all fours, letting your knees and wrists fall directly under your hips and shoulders, try lining up your middle fingers parallel to the long axis of your body and reaching your thumbs toward each other, so they line up on an axis perpendicular to the middle finger.

How does this feel? Do your palms rest on the ground or does your wrist resist bending? (Work on your wrist and finger stretches above.)

Are your fingers relaxed or are some of their joints buckling upward? (Keep working on your finger stretching.)

You might have noticed that in order for you to stretch your thumbs away from your fingers, your entire arm had to rotate. Keeping your hands "in line," bend your elbows to see where they end up. Are they pointing out to the sides? Ideally, our parts move individually, with ease. But as you might have seen, movement of your elbows can be all bound up with rotations at the shoulder. To break up some of this stickiness, rotate your arm bones until your elbows point directly toward your legs—or bring your elbows in as much as you can *without moving your fingers.*

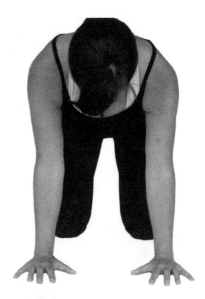

- **Floor Angels**

Another sticky spot restricting the position of the arm bone in the shoulder socket is the anterior (front) chest wall. To mobilize the chest, begin with Floor Angels.

Reclining on a bolster or stacked pillows, reach your arms out to the sides, thumbs toward the floor. Try to get your hands to the floor, keeping your elbows slightly bent. Once your chest can handle this stretch, slowly move your arms toward your head, trying to keep your thumbs on the floor while trying to lift the elbows away from the floor.

This work has two components: the movement toward the floor, and the movement over the head. For some people, the movement toward the floor is easy—not necessarily because they have longer chest muscles, but because their shoulder blades come together more than they should. This is a common cheat that creates what *looks* like shoulder motion, but is actually movement between the scapula and spine. If you notice your arms dropping to the floor with ease, work to keep your shoulder blades apart. It helps to reach your elbows away from you, as though your arms' bones are in traction.

• Windmill Stretch

This Windmill Stretch is similar to a Floor Angel, only the distance traveled by the arm can be greater because it starts farther from the floor.

Start by bringing your left knee up toward your chest. Roll your body to the right, until your knee rests on the ground. Reach your left hand toward the ceiling, moving that shoulder blade away from the spine. Slowly drop your arm to the left (it's okay if it doesn't go to the floor) until you find the boundary of your tension. Once there, pretend your arm is on the face of a clock. Slowly move your arm between twelve and three o'clock, continuously reaching your elbow away from your torso.

I say, "take your elbow away" instead of "take your hand away," as reaching the hand tends to create a motion that straightens the elbow instead of a tractioning motion. Repeat on the other side.

When a rib thruster lies on their back, their ribs will flare, taking their arms and rib cage out of the planes of motion necessary for this exercise. Whenever you do any supine (on your back) exercise, you can make better headway if you bolster the kyphosis in your spine. Sure, your head may be off the floor, but when you're bolstered, the loads to your spine are less damaging and the exercises you do become more effective.

HANGING, SWINGING, AND MORE

Once you've improved the strength and mobility of your upper body with some of these exercises—it can take anywhere from a few weeks to a few years—you can begin to consider hanging and swinging!

Hanging from a bar uses an entirely different set of muscles and motor programs from actually taking your body down a row of monkey bars. It's sort of like the difference between standing and walking. When you're just standing, both legs share the weight of your body; walking requires the muscles on one leg, specifically the muscles of the lateral hip, to be strong enough to hold the entire weight of the body. Hanging on the bars means using two arms to hold your body weight. Swinging from bar to bar or climbing up into something like a tree includes moments where one arm, specifically the lateral muscles of the torso, holds your weight. And actually, once you start moving from bar to bar or limb to limb, the accelerations make the loads to your arm muscles even greater than those created by your resting (read: just hanging there) weight. The moral of this story is: *start with hanging*.

- **Hanging**

Start your hanging journey by applying a very low tensile load, by reaching out to "hang" from a vertical pole. Adjust the distance of your feet and elevation of your hands for varying loads.

Find low horizontal bars (try different heights!) to hang away from, trying both a palm-up and a palm-down grip.

When introducing your first "big" tensile load, find a bar or branch that allows you to keep your feet on the ground. This way, your legs can determine just how much weight you place on all the connective tissues that run between your fingers and spine.

Try both arms first, eventually trying one arm at a time. Until you can hold yourself with both arms, do not progress to single-armed hangs! Notice if your elbows extend excessively—your elbow joint should not go beyond 180°. Work your biceps to stabilize this joint. Do your shoulder blades end up at your ears? See if you can bring them down toward your waistband, giving your shoulder a little more support.

Once you can transfer your weight to your arms, find a bar that allows you to hang with your feet off the ground.

- **Swinging**

Once you've adapted to hanging for longer and longer periods of time, you can start with a two-handed swing.

Swinging from a bar increases the loads created by your body. Adaptations to swinging will come in handy if you start doing one hand at a time.

Before you tackle the monkey bars in the traditional sense—straight on—you can actually "shuffle" down the sides. This movement is similar to the way older people (or anyone without single-leg strength) tend, while walking, to keep both feet on the ground because they can't hold their body weight on one leg very well. You can use this upper-body "shuffle" as a step toward brachiation (the hand-over-hand motion used for traveling down the monkey bars). The scooting motion slightly increases upper-body loading and makes a great intermediate step to play with.

You can also repeat this "shuffle" a different way, with each hand on the long parallel part of the bar. Think *palms facing each other*. This is my favorite way of taking to the bars!

In both these cases, you can slowly add a swing to your torso that makes the shuffling easier. As you get stronger, you will find that you become more adept at manipulating your center of mass (by swinging) so that the momentum of the swing reduces the work necessary. If you ever have a chance to watch monkeys brachiate, you will see this is how they drive a lot of their motion!

• Monkey Bar Triumph

I've heard this story over and over again: "Even as a kid, I wasn't strong enough for the monkey bars!" I have also witnessed these same people making it all the way across the bars for their first time, in their thirties, forties, and yes, seventies.

The easiest way to make it across the bars is moving one hand forward one bar, then bringing the second to that same bar. This reduces the distance (and time) you're supporting yourself with one arm. As you get stronger, you can skip a rung—meaning you never have two hands on the same rung. Once you've done this, please get down and give yourself a huge hug and a fat pat on the back. You're a rock star today. A rock star with monkey arms.

• Rainbow Monkey Bars

One of my favorite stories about monkey barring comes from the parent of two kiddos who were climbing aficionados at their local park. When visiting a new playground, her kids met, for the first time, a monkey bar set that rose to an incline. The first set out with the familiar hand-over-hand motion she had been using for years, only to, oomph, miss a bar as the set rose away from the ground. Her sister missed it too. They scampered around to take a second pass. Boom. Boom. They both fell at exactly the same spot—where the bars did not present what the girls were expecting. By the third pass, the girls stopped the autopilot and changed their geometry to meet the environment—and made it all the way across.

I've listed fixed, straight bars as the fundamental equipment for hanging because they allow you to set a baseline to measure your progress, but there are many options that offer different loads and utilize the body uniquely. Monkey bars designed with inclines offer different cellular deformations. Unstable objects, like rings that swing freely, require you to bring more muscle to the game. If you want some serious work, try climbing up a vertical rope or bar. Tarzan was no joke. Swinging from vine to vine takes some superhuman strength that crossing a set of monkey bars will not prepare you for. There's always some place to go next, if and when you're ready.

• Au Naturel

Here's an analogy for you: Hanging from bars is to hanging from trees as walking in the mall is to walking on natural terrain. To save me some writing time, go back and read everything I wrote in chapter 5 about the additional benefits of walking on natural terrain. Then replace the word "walking" and "terrain" with "hanging" and "surfaces" and you'll catch my drift.

The endless variation of branch angle and girth, bark surface and tree geometry gives you not only a different physical load, but a mental one as well. I can't say this enough: A repetitive environment breeds mindlessness. The human is constantly expending a lot of energy up front to learn, only to put the skill in an automatically run file—no energy (or thought) required. You want to kick some serious health-butt? You've *got* to mix it up.

IT'S HARD TO REST IN A ZOO
CHAPTER 8

While acute stress responses can be considered adaptive, enabling animals to escape from danger, chronic stress responses are fraught with danger to the long-term health of captive animals.

KATHLEEN MORGAN AND CHRIS TROMBORG

In earlier editions of this book, I chose a picture of a woman sitting in front of a fire for the cover. Why? I mean, isn't this a book about movement? Why was she just sitting there, outside? Shouldn't she have been on a walk or something?

But I chose that picture specifically to make a key point stand out: it is not only movement—in the conventional sense—that physically deforms your cells and creates chemical signals. The book cover has changed now, but the principle it pictured remains: As much as you can move things in your environment, your environment can move you right back.

This book, as you know by now, is about loads. Loads can be created by movement, but your cells can also be moved while you just stand there. The contraction of your eye muscles in response to something coming into your field of vision, the deformation of the hair follicles in your ears in response to sound, the pressure (or lack thereof) from an extra-padded sleep surface, and the pinches and cinches of your favorite outfit are all examples of how forces can be created by the environments you select. For these mechanosensors—the ones stimulated simply by being in your habitat—the only way to stop unnatural loads is to physically *move your DNA* out of the environment that is creating them and give your body a much-needed break.

While movement certainly creates an internal environment, the opposite is also true—environment can shape the motion of our cells to a degree that is practically invisible to us, both literally and culturally. Outside, away from the hum of the fridge, the light of the iPad, catapulting cars, and a mile-long to-do list, I imagine the woman on the old cover is experiencing, for the first time in a while, a radically different and much more nutritious load profile.

STRESSORS

Stressors are actual or perceived threats that precipitate stress. The response to a stressor requires a shift in normal biological function in order to cope physically and chemically. Mild stressors shift us away from normal function slightly, and more taxing events shift us for longer. And then there are the mild stressors that are small but continuous—that almost never stop. They keep us slightly left (or right) of normal all of the time.

ANATOMY BIT

Is the thought of getting dressed stressing you out now? Don't worry, I'm not going to tell you to get naked all day long. But I will say that there could be load-altering items in your closet that you wear all day long. Bras and briefs (as I've already mentioned) are the obvious load changers. Tight jeans, belts, "body-shaping" garments, and corsets can impact the pressures in your pelvic and abdominal cavities, affecting digestion, breathing, and who knows what else!

NOT ONLY DO WE DO LESS, WE ALSO DO MORE

The paradox of the modern world is this: Not only do we do less, physically, than ever before, but we also almost never do nothing. Our bodies, deprived of large movements, are inundated with subtle-yet-continuous physical stimulation from noise, light, data, etc. This constant stream of input is a two-fold stressor, as not only is the frequency of certain environmentally induced loads extremely high, but also the types of input we are experiencing are unnatural.

In addition to the direct impact this input has on your physical structure, there is also an indirect response in the form of muscular tension—a static tension not associated with movement. Over time, this clenching the body in response to environmental stressors can leave us tense and in need of some serious relaxation.

. . . .

HYPERTONICITY

In muscle, a chronic shortening in response to mechanical, chemical, or psychological stressors.

. . . .

While I'd be the first person to sign up for a vacation to the Bahamas or a day at the spa, when I say we need to relax, I mean we require the physical release of our tense structures. This is supergood news, because Caribbean cruises cost thousands of dollars, and the stuff I'm about to tell you can be done, whenever, for free.

Though it may be surprising, the remedy for tight muscles is not always "stretch them." In some cases, there is no way to apply a stretch. How do I stretch the lens muscles in my eyes? In other cases, muscles are tense because you are actively contracting them—perhaps for so long that you no longer even recognize that you are—and stretching muscles while you're actively contracting them doesn't lend itself to making much net change. As I described before, sometimes muscles have physically become shorter via the absorption of mass, in response to chronic joint positioning. Stretching, while not without merit, isn't always the path to body restoration.

Another seemingly logical solution for muscle imbalances (a tight muscle on one side of a joint that is preventing its full use) is to add tension to the weaker side. But as I've also already described, balancing an inappropriate tension problem with more tension is not going to help increase mobility or strength, or reduce pain.

Rather than stretching or adding exercises to address our chronic stress, we will focus on *releasing* areas of the body most prone to hypertonicity.

CORE ACTIVATION, SCHMORE ACTIVATION

In the 1980s, the standard exercise-class protocol was two hundred crunches, "keeping your back pressed to the mat." But it turned out that this posterior tucking motion coupled with spinal flexion created a whole world of high load to the discs of the spine. In the 1990s, the standard abdominal-work protocol became to maintain a small curve under the back (i.e., no tucking) as you flexed your spine. This way of stabilizing the pelvis was then deemed the *best* way of using the abdomen (read: saving the spine) while doing a hundred to a thousand repetitions of some sort of abdominal move over the course of an hour-long core strengthening class. By the 2000s, a new spinal stabilizing muscle—the transverse abdominus (TrA)—was the darling of exercise classes, research, and "Can This Core be Saved" magazine articles the world over. And that's a good thing, because no one had ever talked about this muscle before when discussing the abs. A pretty huge oversight.

It was no surprise that over a few years, information about the TrA spread into all areas of musculoskeletal health, including guidelines to "keep your belly button pulled in toward your spine" (no, you don't move your pelvis to do this) *all the time* in order to protect your lower back.

We tend to operate under the assumption that if something like strength or the activation of a particular muscle group is good, more must be better. We act as though the concept of homeo-stasis—well understood with regards to the natural balance necessary for our (and all) biological systems to function—doesn't exist. As if there's no such thing as overdoing it in the body.

Let me explain a bit about your muscles. Muscles—your TrA included—should be able to generate enough force to do their job. Not more. Not less. Just the right amount necessary, depending on the situation. Your TrA is a dynamic muscle that responds to the loads placed upon it. Changing positions, flexing, extending, and rotating the spine, coughing, speaking, speaking louder, laughing, singing, eliminating, birthing, vomiting (my least favorite thing in the world), and walking are all movements that require the TrA to respond appropriately. In every one of the examples I've listed, the TrA is active, but in a unique way. Meaning, the work done by the TrA is less when I'm singing than when I'm running, but more when I'm running *and* singing. The TrA works more during the expulsion phase of giving birth than when I'm doing a plank exercise. Less when I'm lying down compared to when I'm standing, but more than either as I transition between the two.

Movement enthusiasts the world over have been instructed to stabilize their bodies by pulling their belly buttons to their spines, and I'll agree that there is a benefit to having the TrA activate when loads are placed on the body in a way that puts the spine at risk. But the TrA should be activating *on its own*. If you have to think, "Better stabilize my core now," then it means that your muscles are not sensing the load, or your muscles are sensing the load and do not have the leverage to generate

CONT'D ON NEXT PAGE

CONT'D FROM PREVIOUS PAGE

force (see how the addition/reduction of sarcomeres affects force production in chapter 4), or the loads you place on your body are infrequent and higher than the loads you place on it most of the time.

If you spend most of your time not moving, then of course, the infrequent task of moving furniture, sliding heavy pots when gardening, or whatever else you've done to throw your back out will create loads your muscular system is not strong enough to counteract. It seems logical to compensate for this weakness by keeping your belly pulled in all the time to stabilize your spine. But here's the problem: By keeping your belly engaged all the time (at a fixed amount of force that may be more or less than what you'll actually need to complete the task), you're essentially "casting" these muscles to a certain length, reducing their ability to respond appropriately to different situations.

I like to think of the TrA—well, all muscles actually—as a tennis player. A game of tennis is made up of the following shots: the serve, forehand, backhand, volley, slice, smash, and lob. For every one of these shots, the objective is to get the ball over the net. But in order to do this, the tennis player must stand in a way that makes each of these shots an equally viable option. What would happen if a smash was the only shot that could get the ball over the net, but the tennis player was standing in the unique stance for the backhand? The player, preorganized in one position (racquet across the body), would not have enough time to reorganize into a different one (racquet overhead) and would fail the task.

In order to stay ready for whatever type of shot is necessary, the tennis player must stay in a ready position—a position that allows the tennis player to hit all possible shots equally well. Resetting to neutral between shots ensures a player can respond most appropriately over the period of the entire game, and win. In this same way, the geometry of your TrA's sarcomeres must be ready for whatever activity you perform, whether it's a selected activity, like bending over to pick up a child, or involuntary, like coughing. (And, P.S. This goes for your pelvic floor too, which you can read more about in chapter 10.)

When you train your abdomen in all the time, or have habits that alter its structure and force production capabilities, you disable its function. In order to restore the natural function of the systems located in your trunk, you have to let your body run the sophisticated programs located in the brain and not override these programs with the maintenance of a single position (belly button to spine) assumed worthy based on limited perspective. A released gut, muscles with the optimal lengths as determined by your most-utilized ranges of motion, and lots of moving in unique ways will not only assist your spine when needed, but not interfere with all the other "shots" necessary for playing the game of life. Your stomach will flatten *as it needs to*. A continuously flat stomach, while perhaps valuable for cultural purposes, is not necessary for physiological ones.

RELEASE THE ABDOMEN

According to the National Institutes of Health, at least sixty to seventy million people (in the US alone) have a digestive issue and of those, sixty million are chronically constipated—sixty million! In the best cases, people seek alternative nutritional therapy, and there are five million who take some sort of prescription to help the essential human process that is digesting food. I'll bet that in most of these cases, the mechanical environment in which the intestines dwell is never even considered.

If you have a mirror handy, I'd like you to stand up in front of it and strip off all of your clothes. Now, pretend someone is going to snap a picture. What do you do? If you are like most people (or, at least, like me) you will probably do some version of sucking it in. Look, Ma! No belly!

Many people keep their "festiveness" hidden in this way, and some create this inward motion of the belly because they believe that this is how to keep the abdomen toned to support the back. But not only does sucking your gut in *not* stabilize your spine and minimize belly fat in the long run, it can actually worsen both of these issues.

"Sucking it in" would be better described as "sucking it up." Are you still standing in front of that mirror? Let your belly hang all the way out. All. The. Way. Out. Take a look at the mass attached to your abdominal wall. Have you ever wondered where it goes when you bring it in? And, since your abdomen is tightly filled with abdominal stuffs, have you ever wondered where those items—like your organs and intestines—are going all day long?

It's pretty simple. The mass on your belly, when you draw it in, displaces the

ANATOMY BIT

Sucking in the stomach is different from activating your transverse abdominus (TrA) muscles, which assist in stabilizing the spine. One way to tell if you are mistaking tucking the pelvis for activating your TrA is to get on your hands and knees in front of a mirror. After releasing the abdomen, draw your belly button up to the spine without moving the pelvis. If the pelvis does change position, then you have recruited muscles that are not the TrA. You can practice isolating this motion, but in the real world, the TrA is best activated by you moving your body throughout the day. Keeping your pelvis and rib cage neutral to each other will assist the function of these muscles by optimizing their fibers' leverage.

stuff in your abdominal cavity a distance equal to the space occupied by your bellymass. Your subcutaneous (under the skin) fat can be pushing you any combination of up, down, or sideways, but in every case, your guts are now in the territory of other parts and functions. The constant upward motion can displace the diaphragm and eventually push your organs right up into your thoracic cavity—also known as a hiatal hernia. If your organs go south, the extra downward pressure can lead to an inguinal (groin) hernia or create a pelvic floor disorder.

If you've got a lot of belly fat you're trying to hide, or no fat but a serious vanity issue, your chronic hold-in patterns can be wreaking havoc on all sorts of other body parts and functions that no one ever thought to link together.

• Abdominal Release

This release will help you see if you're a gut-clencher. Begin on your hands and knees, in front of a mirror if you have one. Start by relaxing your spine all the way. Watch to see if the pelvis stays tucked. If it does, focus on allowing it to unfurl. Your tailbone should lift up toward the ceiling as your pelvis rotates toward the floor. Now, let your abdominal wall start to drop to the floor, noticing any tendencies to bring it back up to the spine. Keep releasing until it touches the floor. Just kidding. It probably won't touch the floor, but it should feel like it's miles away from you!

Note: Don't *push* your abdomen away from you. Simply relax the muscles that are preventing gravity from taking full effect (which isn't as easy as you'd imagine!).

Once you're all the way out, hang out and breathe slow, calming breaths. Notice if you've got any internal dialogue running through your mind like, "I hate my stomach!" or "My spine is so vulnerable!" This will give you insight into your relationship with this tension habit.

Now that you've experienced the sensation of full abdominal release, you can do this while standing and sitting. Keep checking yourself for gut tension throughout the day, softening when you catch yourself clenching!

My favorite story about the Abdominal Release comes from a physical therapist who gave this exercise to a woman seeing her for chronic constipation. The PT had one session with the patient, who hadn't pooped in over a month (!!!). The woman was able to go after just three days of releasing her abdomen. Why? The movement of food through your intestines can be affected by larger forces created by the muscles of the trunk, diaphragm, and pelvic floor.

ZOOS (AND STRESS)

"Evidence of the damage from these noise factors has only recently come to light. With the advent of the new field of bio-acoustics, patterns are beginning to emerge, through the use of new field research techniques, that confirm the loss those of us particularly sensitive to the natural world have instinctively been feeling for some time. The following examples demonstrate the point:

Many types of frogs and insects vocalize together in a given habitat so that no one individual stands out among the many. This chorus creates a protectively expansive audio performance inhibiting predators from locating any single place from which sound emanates. The synchronized frog voices originate from so many places at once that they appear to be coming from everywhere. However, when the coherent patterns are upset by the sound of a jet plane as it flies within range of the pond, the special frog biophony is broken. In an attempt to reestablish the unified rhythm and chorus, individual frogs momentarily stand out, giving predators like coyotes or owls perfect opportunities to procure a meal. While recording the rare spade foot toads (*Spea intermontanus*) above the north shore of Mono Lake in the Eastern Sierras only a few miles from Yosemite National Park one spring, a similar event actually occurred. After the sound of the passing military jet disappeared, forty-five minutes passed before the toads were able to reestablish their protective chorus. In the fading evening light we observed two coyotes and a great horned owl feeding by the side of the small pond. Because of the unique manner by which we record and measure sound, we have discovered that the relatively intense sound produced by a low-flying jet aircraft can cause changes in the biophony that induce certain creatures to lose the life-saving protection of their vocal choruses."

—Bernie Krause, from his speech "Loss of Natural Soundscape: Global Implications of its Effect on Humans and Other Creatures"

Facilities that house animals are, of course, very different from natural habitat, which makes them great resources for studying the impact of abnormal stimulation on behavior. It comes as no surprise that most of the unambiguous data with respect to biological stress comes from researching animals in captivity.

When compared to original animal habitats, zoos are noisier (and sometimes quieter). The intensity and cycles of light in zoos are problematic for the animals, especially when the quality of light is unnatural (fluorescent). Researchers in this field recognize that it is likely there are unknown but critical components to an environment that promote the development of cognitive and sensory abilities. Nature provides a multitude of variables that are necessary to the animal kingdom.

Animals exposed to these stressors—artificial noise, lighting, scheduling, etc.—demonstrate increased abnormal behavior, self-injury, impaired reproduction, and a suppressed immune response.

ANATOMY BIT

To find your temporal mandibular joint, put one hand on each side of your face, just below your ears, and open and close to feel where the jawbone articulates.

RELEASE THE JAW

There is a myth about muscle—that tension equals strength, and stronger equals better. The jaw makes a great example of how muscle tension can be quite harmful to structures affected by that tension.

I was secretly pleased when my dentist pointed out that while my husband was a teeth-grinder, I was not. (I'm very competitive.) I wasn't pleased at all when she said that I was a jaw-clencher. Oops.

The habit of not just closing your mouth, but also gnashing the teeth together via guy-wire-worthy jaw tension, places an abnormally high load on the teeth, which creates visible signs (warning flags!) of too much tension. These places of strain become weaker points in the teeth, which in turn become the places of future cracks. But the teeth are just one, easier-to-see, end of the tension-creating unit. The temporal mandibular joint (TMJ) is also being inappropriately loaded. For many, this tension habit leads to pain in this area.

- **Jaw Release**

Throughout the day, see if you can relax the jaw without opening the mouth. There's a point where the jaw muscles exert more force than necessary. When you catch yourself tensing, let the bottom teeth move away from the top. There. That's better, for your teeth and for your TMJ!

RELEASE THE EYES

I wear contacts now, but when I was a kid I wore thick glasses. Coke bottles. And in addition to being thick, they were huge, thanks to the eighties. I mean, I was wearing these huge, Coke-bottle glasses in the eighties, but my mom, who picked them out, was from the seventies. Really, she was from the fifties, but she was a "child" of the sixties, which means that by the time she was a parent (in the seventies) she preferred lenses that covered the surface area of a small pond.

See?

The apple doesn't fall far from the tree.

I started seeing a new eye doctor last year, and when he was checking my prescription he said—after the obligatory, "Whoa! That's some serious myopia!"—"You must be an engineer or a science-type." What? I immediately pressed him for his technique at assessing career paths from a prescription, and he said that, in general, people who spend a lot of time with their noses in a book tend to have greater cases of myopia. I wonder if this is why the stereotypical nerdy kid in school (guilty) is always wearing glasses?

Myopia has risen to epidemic proportions in Asia, a rise long blamed on book reading or other "near work" like using computers and tablets. While the genes for nearsightedness are certainly at play, myopia is extremely rare in hunter-gatherer populations, leading researchers to look for more specific environmental factors that reduce our ability to see far away.

Research shows that cultures historically without books show an increase in nearsightedness once reading or other hallmarks of a "civilized" environment (like night-lighting) are introduced. But more recent research specifically looking at reading shows that time spent outdoors might be the factor worth looking at instead.

When looking at things close up, the ciliary muscles in our eyes contract. But whether we're reading or not, most of the objects we look at all day long are indoors and still relatively close to us. For home dwellers, this means that even when you aren't looking at your computer or book, the farthest object from your face is, what, twenty feet away? The natural loads to the eyes include looking over many different distances. We've got supernear and kind-of-near covered, but to really load the eyes in a different way, we must look at objects really far away—a process that facilitates the relaxation of tense eye muscles.

So yes, my mom was probably right when she said I needed to put the book down, but she should have also told me go outside. Which, now that I think of it, she did tell me, only I just went outside and started reading again. Meaning the eyes need to step away from your book/computer/smart phone and go outside—they need to be looking at things at great distances to move fully.

- **Eye Release**

The mechanical loads you are missing in your eyes come when you point your eyes at things really far away and let them focus. Cross-train them when you walk! Keeping your head still, look right and left, up and down, and out of the corners of your eyes—all while looking far away!

For those of you stuck at a desk, arrange your office so that you can look out a window a few times each hour, giving yourself a couple of minutes to focus on the farthest thing. Similarly to when you were releasing your abdomen, you might have to try to find where in your eye you are holding tension. Slip off your glasses or contacts when you can, as wearing your corrective lenses makes finding the tension in your eye a little more challenging.

RELEASE THE EARS

> One of the single most important resources of the natural world is its voice—or natural soundscape. **BERNIE KRAUSE**

When I was in grad school, a friend mentioned a website that streamed real-time audio-video from an African waterhole. One night while studying, I popped it open (procrastination is a problem for me) only to see warthogs rooting around the pond in the moonlight! I was totally hooked. I started keeping a window open all the time and became an official "pondie," texting and emailing friends at five o'clock in the morning to say: "ELEPHANTS!" But after the excitement of animal spotting wore off, I kept finding myself turning the pond on for the soothing, low-grade bug noise.

. . . .

BIOPHONY

All non-human sounds produced by living organisms in a given environment.

. . . .

Noise is a continuous phenomenon. With the rare exception of being in a sensory deprivation chamber, your mind is constantly occupied with the task of processing and interpreting sound. In fact, you are so used to hearing all the time that you probably don't even pay attention to how much energy you expend reacting to ambient noise. Distress in zoo animals has been associated with both the continuous, unnaturally frequent and loud building sounds they're exposed to as well as the absence of an environment's natural sounds—an ecosystem's biophony—that play a role in the success of a species. Not only in zoos, indirect man-made noise (the sounds of jets are just about everywhere) now affects the entire globe. As anthropophony overtakes an environment, biophony responsible for reproductive and survival-associated communication is corrupted.

ANTHROPOPHONY

Man-made noise, such as sounds produced by humans both directly (language, music) or indirectly, such as random signals generated by electromechanical means.

Sound experts measuring ambient noise levels in natural settings have found the rainforest to be the most noisy, riparian zones to be in the middle, and savannah habitats to be the quietest of the lot. Rustling vegetation, wind, water, birds, and insects are the world's natural noisemakers, with the peak decibel (dB) of all ambient noise to be about 36 dB midday, rising from an early morning 20 dB.

There is no evidence that constant noise results in death (although there is evidence that if it's really loud, noise can kill your hearing cells!), but there is evidence that noise can reduce your ability to sleep, increase muscle tension, and, in extreme cases, change your body's geometry.

In experiments, humans all respond the same way to loud and sudden noises: eye blink, facial grimace, a bending of the knees, and a flexion (curling-up motion) of the spine. **JAMES DAVID MILLER, EFFECTS OF NOISE ON PEOPLE**

• Ear Release

In short, make your life less noisy. Take walks—at least sometimes—without the smartphone. Drive without the radio. Unplug (or un-battery) any non-essential thing. Invest in some earplugs if you're right up on a roadway or for when you travel.

And don't forget to let the natural sounds in! Open the doors and windows. Take a walk just to work your ears in a different way. Since there is so much about nature we don't know, cover your bets by encouraging your ear's relationship with nature.

NOISE POLLUTION, A MODERN "PLAGUE ON YOUR HOUSE!"

Imagine the world ten thousand years ago or so, when aside from the sound of two rocks banging together as someone tried to make a spearhead, all sounds were made by the natural world. You paid attention to every sound because it provided important cues for your daily survival. If you heard the thump of hailstones, it was probably not a good day to go out scavenging for food. If the big, hairy thing you'd never seen before was making a very loud low, rumbling sound as you approached, you should probably have a more coherent plan than going over and licking it to see if you could eat it. And if your early-human spouse was screaming at you after a long day's hunt, you should probably wipe the mammoth dung off your feet before entering the cave.

As we spread over the planet and upped our technological game, anthropogenic, or human-generated, sounds started becoming the dominant source of noise and signals in our environment, starting innocently with bull roarers and other musical instruments, spreading to the clattering of vehicles (whether horse-drawn or propelled by internal combustion), spreading to ever-on radios and televisions, until the entire infrastructure of our cities became giant low-fidelity speakers, generating noise bands so intense that urban birds and other wildlife had to adapt their mating calls or move out to dwindling quieter areas. And all the time, our brains have been trying to adapt, telling us that 75 dB, which would have signaled an approaching avalanche or earthquake back in the day, is a completely reasonable level for a city street, even if it means we have to put on noise-canceling earphones so we can hear our personal musical soundscape or phone calls above the din.

Humans, like most other social animals, are noisy. The difference is that our technology is too, whether intentionally, like the massive subwoofer-laden car that plays rap at 120 dB as it cruises down the street, or parasitically, like the oven range fan that's purring away at 80 dB to keep the smoke away from the fire alarm that will shriek at you at 100 dB as it criticizes your cooking skills. We pay lip service to trying to control noise pollution by putting volume limiters on children's headsets (maxing out at 85 dB), mufflers on cars (dropping the 120 dB raw sound to a maximum of 95 dB), and zoning regulations around airports (supposedly limiting residential areas to 65 dB). But we often get tricked by our own brains, whose attentional systems often let us immerse ourselves in excessively loud movies (up to 100 dB), music (often 100 dB or more), phone calls (with ringers at 85 dB and call loudness up to 90 dB), and other media by making us think that if it's really too loud, we'll know it and we'll just turn down the volume. The problem is that your perception of "too loud" can take so long to kick in that not only do you subject yourself to chronic stress, but also you've already damaged your hearing. And eventually, some years later, when the silent damage has crept up on you, you find suddenly that everyone is mumbling, cars "sneak up on you," and your inability to hear the relatively quiet "normal" sounds of life leave you paranoid and accident prone.

CONT'D ON NEXT PAGE

CONT'D FROM PREVIOUS PAGE

Hearing is a sense that is hundreds of millions of years old; it works in the dark and out of the line of sight and even works when you sleep. It is an alarm system and a gift for communication bestowed on every vertebrate on the planet. But it is having trouble keeping up with our technology, which throws more and louder sonic information at us every day. The best way to save your hearing is to pay attention and realize that if you think it's too loud, it's *way* too loud.

—Sidebar by Seth S. Horowitz, PhD, auditory neuroscientist, CEO of neuropop.com, and author of *The Universal Sense: How Hearing Shapes the Mind*

RELEASE THE PSOAS

The psoas (SO-az) is an unfortunately named muscle directly impacting more joints of your body than any other single muscle. Branching off like a crazy starfish, the psoas muscles (you've got a set of them, one on each side of your spine) attach to all the vertebrae between the bottom of the rib cage and your sacrum and the lumbar spine's intervertebral discs before terminating at the top of your thigh bone. A "sticky" psoas has the ability to alter the orientation of your spine, hips, pelvis, or knee—and potentially all four of these parts at once.

Embedded between the layers of the psoas is the lumbar plexus, a dense collection of nerves that innervates the abdominal muscles, the pelvic floor, the deep hip rotators, and most of your thigh muscles.

Because the psoas attaches at multiple locations, passes over multiple joints, and entraps a major neurological network, it is no wonder that so many injuries can be blamed on one misbehaving muscle.

• Psoas Release

Is your psoas too short for your body height? Here's both an assessment and the release.

Place a bolster, sleeping bag, or stack of pillows behind you. Start by sitting on the floor with legs extended. Relax the muscles of the thighs until the hamstrings (muscles on the back of the thigh) rest on the ground. You might need to untuck your pelvis for this to happen. (Note: If you had to untuck, that's one sign of a short psoas.)

Once your thighs are down, start to recline, stopping just before the hamstrings lift away from the ground. At this angle, bolster your head and scapulae, leaving space for the ribs to lower to the floor. The height of the ribs from the floor is an indication of your habitual psoas tension.

Once you have identified and bolstered your body to meet your psoas tension, you can start relaxing the ribs to the floor. As with all the other "releasing" exercises, the point is not to get your rib cage to the floor by flexing your muscles, but to lower it down by releasing the ones holding it up. As you release your psoas, your ribs will be able to move closer to the floor, so continuously adjusting the height or position of the bolster will be necessary as you improve.

Releasing your psoas can take anywhere from ten minutes to ten years. It depends on how shortened your psoas has become and for how long, how much you sit when you aren't releasing, and how often you keep your ribs in check. Rib-lifting when standing is a psoas shortener, so being mindful of rib position can facilitate a better outcome over time.

RELEASE THE SPINE

A spinal twist is a pretty common exercise to find in yoga or back-health classes. But I've seen how most people "twist" their spines, and I'm compelled to chime in.

A "twist," by definition, is the *rotation* of one part relative to another. A spinal twist, then, is one in which each vertebra of the spine rotates a small amount to create a net motion that places your rib cage in a different plane from your pelvis.

PSOAS: WHAT'S IN A NAME?

One of the most interesting things about the psoas is the history of its name. Long before Hippocrates began using the modern Latin term *psoa*—muscle of the loins—ancient Greek anatomists called these muscles the "womb of the kidneys," due to their proximity to those organs. In the 1600s, the French anatomist Riolanus committed a grammatical error that survives to this day, identifying the two psoa muscles as the psoas (if there's two, just add an s!) instead of using the proper Latin *psoai*—a goof that would, in the end, perhaps influence our tendency to think of these muscles as a co-functioning team rather than as individual muscles that can adapt to our asymmetrical habits.

Adding insult to injury, in the 1950s John Basmajian, MD, father of electromyograph (EMG) science, argued that the iliacus and psoas muscles could not be expected to have unique functions because of their shared lower attachment site. His opinion triggered widespread use of the term *iliopsoas*, stripping each muscle of its individual identity and setting the precedent of misattributing iliacus EMG measures to the deeper and more-difficult-to-read psoas muscle. All this historical context makes it much easier to understand why the psoas is omitted from most anatomy books, and replaced incorrectly with iliopsoas.

The spinal twists I've witnessed, however, are better called spinal *rolls*. In the case of a spinal roll, the entire spine (and torso) rotates as one, until the pelvis and ribs are in a different plane from where they started, while the arm reaches back to where everything came from, in the hopes of placing a twist on the spine. This roll move has value, but it differs from a spinal release (a real spinal twist) in that rolling the spine doesn't create twisting loads between the vertebrae—a necessary step in restoring the vertebral column's twisting action. And, more importantly, the inability to twist speaks volumes about the tension, again, in the abdomen.

There is hardly any task in the modern world that calls for the rotation of the spine, save looking over your shoulder when you're backing up your car. Rotating the spine while you're loading the rotating muscles—think of ripping out vine-like roots from your garden, or the one-sided pulling in of a long rope—has nearly vanished, unless this motion is part of your work or workout. So it's no wonder that the trunk has adapted by stiffening its rotators.

• **Spinal Twist**

This twist (er, release) is just like you imagine. You're going to start by lying on your back. Bring a knee in, and rotate your pelvis to lower that knee to the opposite side of your body. But there are a few guidelines here that make this exercise different from a twist you might have encountered before.

Difference 1: You must eliminate the thrust of the rib cage before beginning a twist. This means the head, neck, and shoulders should be bolstered with a blanket or two (or seventeen) to keep the rib cage in line with the neutral pelvis. This eliminates the shear and places the vertebrae and musculature in an orientation best suited for torsional loading.

Difference 2: The goal isn't to force your knee to the floor. The goal is to note where your pelvis stopped moving due to tension in the trunk muscles. Twist only as far as you can without taking the ribs with you, no forcing it. This ensures that you'll stay on the edge of the boundary set by your current abdominal muscle tension.

If you find that your pelvis barely moves—that your trunk is that tight—and your knee is nowhere near the floor, you might want to stack pillows so that the knee crossing over can rest on them. This will reduce the load to the spine and keep these muscles from tensing unnecessarily. As you did in the Psoas Release, you can adjust the height of these pillows with time.

RELEASE YOUR PILLOW

This wouldn't be much of a chapter on rest if I didn't talk about sleeping. Sleeping is one of those things that everyone—fit and unfit—does that can really level the playing field when it comes to health.

If you've ever gone camping with a group of adults not accustomed to sleeping on the ground, you're likely to hear the moans and groans the next morning about stiff backs, sore necks, and wonky shoulders. And come to think of it, I've heard these same complaints from people emerging from hotel and guest bedrooms the world over. "That bed was too soft!" "That pillow was too high!"

Mattress and pillow manufacturers present the idea that there is an ergonomically "best" way to sleep and that through the use of these devices, we can get better sleep and thus improve our health.

But your pillow and mattress are subversive immobilizing devices issued at birth.

Don't get me wrong, I love a great mattress in the same way I love a supercomfy couch. I wouldn't argue that these items don't feel good when you lie on them, but "This feels good!" isn't a great way of selecting items to improve your health. I'll bet if you thought about it, you could make a list of items that make you feel good but, quite directly, worsen your physiology.

I've always suffered from headaches. About ten years ago, while stretching my neck to deal with the always-present tension there, I began wondering where and when natural neck stretches would occur. I realized that my ever-present pillow was, in fact, preventing the very motion of my neck I was practicing during my "stretch time." Of course! The six to ten hours I spent in bed every night would require different joint deformations and create different pressures were it not for the pillow and mattress preventing these loads. I resolved to get rid of my pillow.

Getting rid of your pillow is similar to getting rid of your shoes. Yes, you can throw both of these things out the door in an instant, but by doing so you might load underused tissues to the point of damage (as demonstrated by camping-neck or -back). Instead, consider a slow and systematic approach. Pillowless sleeping requires relaxed neck and shoulder muscles. Sleeping without a pillow requires less hyperkyphosis (the excessive curvature of the upper back). You can work on these mobilities during your awake time.

It took me about a year to go completely pillow-free, but what I did was change my pillow height over time. I went from something big and fluffy to something medium and fluffy. I progressed to a less fluffy pillow to a towel to a wadded T-shirt to nothing. I no longer need a fixed head position to keep from aching the next morning. By progressing, I gradually loaded the tissues in my upper body so that they could adapt to my "sleeping" workout.

RELEASE YOUR MATTRESS

Did I already say I love my mattress? Well, I do. Or did, actually. About a year ago I decided that I had to start letting this other item go. My new role as full-time mother of two had cut into my spare time, which meant less time for structured movement (read: exercise sessions). My body didn't feel good anymore, and I had to take joint mobilizing where I could fit it in, which turned out to be *while I was sleeping*.

My progression out of my most favorite mattress began organically with the bed-swapping we did as a young family. (Something they don't tell people before they have kids is, "You will probably find yourself sleeping in different beds, in different rooms, in different positions, for different hours than you are used to.")

The different mattress surfaces changed the loads to my body, creating a cross-training effect. Yes, it turns out you don't even have to be awake to cross-train.

After a few months, I no longer needed the mattress to keep me from hurting, and eventually I got to the place where, after sleeping all night on a mattress or pillow, I was as stiff and sore as I used to get camping!

Research into sleeping positions worldwide shows that we supercomfy bed users are outliers in both the animal kingdom as well as human populations. Sleeping in nature—something humans have been doing since day one—creates varying loads in your cells. Right now, your pillow and mattress serve the same purpose as an orthotic or "supportive" footwear. Just as an orthotic supports the weakness created by wearing shoes, the pillow reinforces the body position created by using a pillow. Just as constant shoe-wearing and flat, unvarying terrain have left you with poor foot mobility and strength, always sleeping on something flat and squishy has

ANATOMY BIT

If a tree falls in a forest and no one is around to hear it, does it make a sound? If your body feels great until someone touches it, are you really in great shape? If you've ever gone to a massage therapist only to have them "discover" sore spots on your body, ponder this: Those spots have been sore for a very long time. The question is, why didn't you know about them?

Sleeping on a mattress dampens the pressure to your body. The cumulative effect of sleeping every night on a giant squishy surface is that it keeps you blind to the fact that parts of your body have become so immobile and inflamed that being touched hurts. If you've gone to the massage therapist and found relief from the application of pressure over time, consider the compound effect of getting "worked on" by the floor every night!

altered the mobility and sensitivity of your parts. The joint alterations required for ground-sleeping are natural and they're currently underused.

There's not really a problem with your pillow or mattress or a "bad" position to sleep in. Again, it's the repetitive nature with which we do everything that atrophies the tissues necessary to do something different.

RELEASE YOURSELF (FROM THE INDOORS)

I've always been a tree-hugger, because trees are valuable. And they're not only valuable because I *believe* them to be; trees make the very oxygen that you and I need to survive. Trees and humans are in a continuous relationship of gas exchange. They need us and we need them. But a tree's value to our health isn't only that they make oxygen.

Shinrin-yoku, or "forest-bathing," is the process of making contact with and taking in the atmosphere of the forest. Heavily researched in Japan, forest-bathing has been shown to promote lower concentrations of cortisol, lower pulse rate and blood pressure, and a reduction of "technostress," as measured by a reduction in cerebral activity. While anyone who has spent a few hours wandering out in nature can tell you that yes, it's very relaxing, it is through scientific investigations that the mechanism behind our physiological response to the trees is better understood. We aren't responding to a tree per se, but rather undergoing an invisible interaction with phytoncides—active chemical substances given off by plants. The tree, secreting these substances to ward off harmful bugs and rot, is also providing us with a compound that does our body good.

I've said it before and I'll say it again: nature is complex. In the same way it is impossible to list all the benefits of eating a whole-food diet, it is impossible know every single exchange going on between us and "being outside." In our desire to simplify nature into the nuts and bolts we can duplicate while living separate from nature, we miss the obvious picture: part of the thing is not the thing.

Vitamin D is one such nut. Interactions between us and the sun, it turns out, are necessary to produce certain types of vitamin D. Areas that are low in light (or are adequate in light, but have a population that just doesn't go outside very much) have higher rates of illnesses stemming from low light. But, after further investigation, it turns out that vitamin D isn't the only beneficial nut of sunlight; there's also the bolt of the exposure to ultraviolet (UV) radiation. Independent of vitamin D synthesis, UV radiation decreases risk factors for inflammatory demyelinating diseases like MS. In short, you can't take vitamin D and get the full benefit of the sun. Or simply take a UV shower. We haven't even begun to learn all the physiological processes that depend on interacting with the sun. Or the wind, the warm, the cold, the ground, or biophony, for that matter.

Science does a great job at reducing variables until they are small enough to be understood, but we aren't doing a great job at reassembling the picture once it has been broken down into a thousand pieces. The biochemist researching plant-secreted compounds doesn't talk to the physiologist researching the electron transfer between the earth and the bare foot (or "grounding"). The immunologist researching immunity-boosting parasites in the dirt (that would have been on our food if we weren't so obsessed with washing it all off) doesn't talk to the anatomist studying the activity of millions of tiny muscles—called *erector pili*—that contract to lift your body hairs when you get goosebumps in chilly weather.

Safe and warm in our homes, offices, cars, and clothes, we have missed out on an impossible-to-quantify number of biological interactions, and our physiology is the worse for it. While we are extremely lucky to be living in a time when synthetic materials are constantly in production in factories the world over to offset our lifestyle choices, these materials do not duplicate nature. They are simply spot-treatments that come with a tax on humanity and ecology. Outdoor time, like movement, is not really optional—your body and community truly depend on it.

REST, IN GENERAL

If exercise is almost universally listed as a disease reducer, "stress" could be listed as a disease risk factor. The difficulty with studying stress is pinpointing where, exactly, the (perfectly natural) occurrence of stress results in physical distress.

Everyone copes with stress differently—our ability to do so changes with age, experience, and even our current state of health. There are areas where stress is unavoidable, but as this chapter points out, there are many areas in your everyday life where it can be significantly reduced by simply changing your environment.

Most of the chronic physical deformations listed in this chapter will not take additional time to undo. Turning off the radio doesn't add time to your day. Sleeping on a few blankets instead of your mattress doesn't need to be put on your schedule. When we start to evaluate motion on the microscopic level, we begin to understand how much *easier* health can be. The notion that we must work intensely (read: do a lot of exercise) to stay healthy inside our zoo-like environment is not necessarily invalid, but it implies that there is nothing we can do about our captivity or that we've agreed to staying in our cage of comforts. Behavioral recommendations stemming from this point of view fail to recognize that we ourselves have the key to unlock the door and move our DNA to a new environment *whenever we choose to*. What are you waiting for?

WALKING: THE SPECIFICS
CHAPTER 9

n the 1980s, Dutch biomechanist Gerrit Jan van Ingen Schenau wrote his PhD thesis on speed skating—specifically, exploring whether a speed skater's performance would benefit from a *klapschaats*, an ice skate that had a front-hinging blade (as compared to the fully attached blade found on regular skates). He predicted that a blade that detached in the back would create a change in the leverage of the calf muscles, improving the performance of competitive skaters.

He was right. In 1994, he had a group of junior competitive skaters wear the modified skates and measured their performance at various points in the season. By the end, skaters using the "clap skate" improved their performance an average of 6.2 percent—compared to the 2.5 percent improvement found in tradition-al-skate users. Once the test group started placing at national events, word got out and higher-level Dutch skaters started using them. At the 1998 Winter Olym-pics, world records were shattered by those wearing the clap skate. The sport was forever changed. So why and how did the skates work?

Van Ingen Schenau originally thought that the skate would allow for greater calf-muscle usage, thus improving performance, but later research clarified that the hinge on the skate changed the lever system not just for the ankle (which yes, allowed for more calf usage) but for the *entire body*. By moving the point of rotation from the toe of the skate to the hinge of the blade, a skater's push-off lever is not just the length of their leg, it's their leg length plus the extra distance to the hinge. A longer "leg" meant a skater's pushing phase—the distance between the foot being placed down and that same foot coming up again—became longer in length as well.

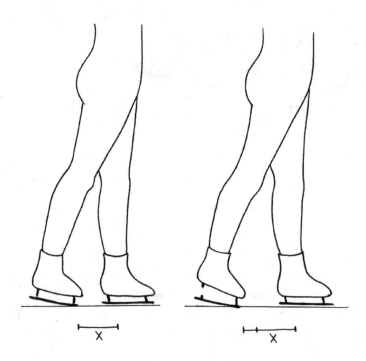

This was important to skaters, as they can only *push* their body forward when their foot is pressing into the ice. Once a skate leaves the ground, the skaters are coasting on the work done by their pushing muscles up until that point. For speed skaters, the longer in distance the pushing phase can cover, the better chance they have at accelerating and beating their time.

This is how walking works, too. Not because we're trying to go as fast as we can when we walk, but because the amount of controlled, forward movement per step depends on the distance our active push-off phase covers.

Like the clap skate, your body has many ways to increase the length of your "leg" and thus the push-off phase. Your maximal push-off could be said to be the sum of hip extension (moving your leg back at the hip joint), spinal extension (moving the hip joint back by arching the lumbar spine), pelvic rotation (twisting at the waist to move the hip joint behind you), and ankle plantarflexion (pointing the toes).

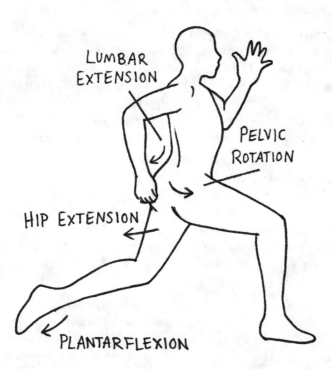

In the case of a rapid getaway or burst of motion necessary to capture something for dinner, you would want to call on all of these joint deformations, as together they increase your leverage and ability to rapidly accomplish your task.

TAKING A CUE FROM THE ANIMAL KINGDOM

To see this whole-body recruitment in action, check out the side view of a cheetah running to catch some prey. You will see the ankle, knee, hip, and spinal hinges flexing and extending their full amount to maximize the body's performance over a short period of time. It's truly amazing to witness. (And if you don't have any local cheetahs to watch, check out YouTube.)

TREADMILL WALKING: AN ENTIRELY DIFFERENT (INTERNAL) ENVIRONMENT

In both professional and layperson contexts, I have seen the treadmill referred to as a technology that changes the *location* of walking or running to indoors, but nothing about how it changes the act of walking itself.

In overground walking, the sole of the foot (or shoe) interacts with the fixed ground, and via this physical connection, fires a specific pattern of muscles to push down and back against the ground, creating a specific ground reaction force that in turn moves the body forward and over the leg that is working in the backward direction. In treadmill walking, the belt is already moving back, which limits the leg's ability to actively push back (more on why this is important in chapter 10). On a treadmill, the legs work in the forward direction to "catch up" to the belt's backward motion, so as to not get pushed off the back of the machine.

At a glance, there doesn't appear to be that much difference between the two. The limbs are moving in the same basic ways—legs and arms swing front and back relative to the body, covering about the same distances. Metabolically, they both burn about the same amount of calories (with the treadmill sometimes coming in a little higher). Research comparing the two has found some biomechanical differences in cadence, stride lengths, the amount of time one is on "one foot" while walking, and the distance covered by the hip joint. Treadmills tend to require more hip flexion, which makes sense since the amount of hip extension is limited.

When it comes to fitness outcomes—burning calories, improving cardiovascular performance, reducing stress—the differences between the two are probably moot. But when it comes to the way our structure adapts to loads—specifically, the skeletal and muscle shape that results from the particular use of the muscles working—the difference between the two is tangible.

I've seen it written that, "Ultimately, your body doesn't know whether you are on a treadmill or a trail," but I beg to differ. I'd say the similarity between overground and treadmill walking depends on your perspective. If you're comparing a few isolated variables at a time, you can get them to look similar on paper. But the body (and its trillions of mechanosensors) does not adapt based on our variable isolation. Instead, it sculpts an entirely different picture (read: body shape) when it's on the ground vs. when on a treadmill.

After watching a cheetah sprint, though, find a video of a cheetah *walking*. What you will see is that the cheetah does not always hinge to the full extent of its capabilities, but rather mostly uses the hip as the forward-pushing lever system. Yes, the cheetah *could* undulate its spine, use its cheetah calves to spring forward, and change the recruitment order of its front legs with every step, but it doesn't. Why? Because moving in that way uses a tremendous amount of energy, both in muscle force and tissue expenditure. The cost of moving that way is very high.

The genetically constituted architecture of bones and joints of the body is based on historical use patterns. If cheetahs always walked by flexing and extending at the spine, I hypothesize that their vertebral joints would have evolved to become equally or more robust than the joints at the hips and knees. Instead, its bony shape, like ours, reflects the way the cheetah has used its body over thousands of years. Cheetahs pad around much more often than they sprint, and the size and shape of their joints match this use pattern.

The human hip is an extremely robust joint. Long levers and huge muscle mass set the forward-propelling capability of this joint very high, yet most modern humans barely use the hip joint when walking. This is a tough concept to understand, because if you take a look around, people's legs seem to be moving forward and back when they walk. If their leg is moving relative to the ground, doesn't this mean that their hip is working? Well, no.

Since we have spent so much time sitting, and because so much of our exercise time is also spent in hip flexion, the mass creating the length of the muscles connecting the thigh to the pelvis (the iliacus) and the thigh to the trunk (the psoas) has decreased. Most of us no longer have the necessary muscle length (in the rectus femoris, psoas, and iliacus) or strength (in the glutes and hamstrings) to move each thigh behind us with ease.

. . . .

RECTUS FEMORIS

One of the four muscles that make up the quadriceps (front of the thigh) group. It is the only muscle in the group that connects the pelvis to the lower leg. The other three quadriceps muscles run between the thigh bone and the lower leg.

. . . .

When you combine this fundamental hip problem with weak feet, chronically flexed knees, and poor core strength, our walking has become a series of complex falls masking a multitude of tiny movements—movements that do not contribute to our net forward movement, and actively wear through our body parts at an accelerated rate.

THE SAME, BUT DIFFERENT

Despite sharing a similar cultural experience, our specific experiences within the culture are different. While most of us sit in equal amounts, not everyone plays the same sports (or sports at all), does the same type of exercise, or has the same anthropometric dimensions. Which means we all cope with our hip-flexor shortcomings in different ways when we're walking—resulting in a unique-to-you combination of toe-ing off (also called toe-walking), bending and straightening the knee, twisting the hips, twisting the foot at the ankle, rotating the entire body around a polar axis, and hinging above the hip, at the spine.

If you've ever been told you have a "spring in your step," chances are you're a toe-walker. Toe-walkers have great elasticity in their lower leg, and their wasted motion is in the up-and-down ping-ponging that contributes little to forward motion. To get a sense of knee-bending, stand in place and bend and straighten both knees a few times. Now, do that again, only while walking. As you can imagine, it is not necessary for your knees to excessively lower you to and from the ground while walking on level ground.

If you're a Latin-dance aficionado, you're familiar with the necessary twisting of your hips. This motion is great while grooving to your favorite cumbia, but if it's happening with each step, it's an indication that your hip joints aren't working very well. The twisting of the foot at the ankle is similar to the gait pattern of a sea turtle. Just as a turtle's flipper rotates as it pushes back, the foot can act like it's got a pin holding it to the ground; where the front of the foot stays connected to the ground while the heel rotates toward the midline. Check your socks. Holes under the balls of the feet? You might be spinning on your feet as you are walking.

A side-to-side sway is the first thing that novice gait-assessors will see when observing someone rotating about a polar axis. In a standing and walking human, the polar axis should run perpendicular to the ground, but when your body is so stiff that major joints no longer articulate, one can move forward by leaning to one side and flinging the other side out in front of them. In this way, the polar axis of the body rocks side to side as the non-weight-bearing half orbits around it to catch up. For a better visual, search "penguins walking" on the internet. This way of walking is colloquially referred to as a waddle, but I don't care for that term as it is often associated with excessive weight. You are just as likely to see this way of walking performed by a very lean person.

Hinging at the lower back can be the most difficult to spot because it looks very similar to hip extension. Only, instead of your leg moving back at the hip, the leg and the pelvis move as one piece. The lumbar spine axis is only five or six inches away from the hip axis, but which muscles would you rather be using to walk? The teeny-tiny muscles of your spine or the hunks of flesh (read: butt muscles) sitting just below?

To help you learn to identify different gait habits I've isolated each variable for clarity, but your gait pattern is probably the sum total of all of these, with one or two being the most used.

As you can probably imagine by now, each of these pattern components comes with its own set of overuse injuries. Not that these motions are inherently harmful to the human body, but their frequency of use does not match the body's architecture well. Here's the thing: In addition to overuse injuries based on our own individual gait particulars, we all tend to share the issues that arise from not using the hip joint—something nearly all of us have in common.

The gait correctives listed here really do apply to most people. While we are all anatomically unique and have individual experiences and therefore adaptations, we have only so many ways to compensate for hips that don't extend. The following series of exercises are designed to increase the distance over which *the hip* can carry and control the weight of the body, while minimizing the typically overused articulations of the spine and knees.

- ### Hip Extension Test

To check out your hip's current range of extension, lie on the floor face down, keeping the three points of your pelvis (the right and left anterior superior iliac spines and pubic symphysis) on the ground. Paying extra attention to the pubic bone, making sure that it doesn't leave the ground, see how high you can lift your knee while keeping your leg straight. Did you get very high? Did you have to use muscular force to tuck your pelvis into the ground in order to lift your leg? This abdominal tension is a compensation used to oppose the tension in the hip flexors, but it's not optimal. Ideally, your hip would have the mobility to extend without this coping mechanism, and when you walk, this lack of mobility can mean increased spinal extension in lieu of hip extension.

IMPROVING HIP EXTENSION

The most common suggestion for improving the backwards motion of the leg seems to be, "Get a stronger butt." And let me be the first to say I have no problem with you getting a stronger butt. No one is championing the development of your backside more than I. But to strengthen your butt to match the resistance created by your own muscle adaptations is to fight yourself all day long. Strength will come when you are constantly cycling between releasing the semi-permanent tension in your psoas, quads, and hip flexors, and exposing your hip and butt muscles to more frequent walking. This not only promotes a sustainable relationship between parts (remember the definition of alignment?) but prevents us from training to the point of excessive tensions that wreak havoc elsewhere. A strong butt is great. A butt strong enough to stabilize the sacrum and pelvic floor all while carrying us up the side of a hill is more than great, it's amazing. A butt muscle trained in isolation (think one hundred leg lifts while on your hands and knees, or doing all your glute exercises while lying on your stomach, without a load on the pelvic floor) can be problematic. So, all of this to say, let's release the hips first and allow the glutes to work…naturally.

As I listed above, there are three muscle groups that can keep the leg from swinging behind the pelvis. The psoas release is covered in chapter 8, and the following will cover the other hip flexors.

• Iliacus Release

The iliacus is a muscle that runs between the pelvis and the femur. It shortens to accommodate a sitting position. If you sit a lot, chances are this bugger has adapted. To release, begin with a gentle tensile load.

Lie on your back (bolstering your neck and shoulders as necessary) and prop the inferior (closer to your legs) half of the pelvis up on a bolster or yoga block. (Top image, opposite page.) The superior (closer to your head) half of your pelvis should tip toward the floor, creating a passive hip extension. (Bottom image, opposite page.)

Don't work to rotate the pelvis; this defeats the point of "release." Just allow gravity to create this hip extension for you. Hang out here as long as you'd like, knowing that even if your pelvis doesn't budge, gravity is still creating the forces necessary to signal, "Lengthen!" to these muscles.

You can even do this right now and continue reading while doing so, yes? Seriously. Go do it right now.

- ## Passive Prone Hip Extension

I wrote a lot of this book while lying on my stomach—a position that can just kill your lower back, right?

One of my tricks to maintaining this position pain-free is to prop my pelvis up on a blanket (or I use a half foam roller) so that my ASIS are supported but my pubic symphysis has access to the ground.

By resting in this way, my writing time (hours and hours) became hours and hours of applying a tensile load to the muscles between my pelvis and femur. I, like you, don't have a lot of extra time. Try this while watching TV, hanging out at home, or working on your computer.

• The Lunge

Get down onto your hands and knees (preferably on something soft, like carpet or a yoga mat) and step forward with your right leg until your right shin is vertical. Now, line up your pelvis so that your pelvis is in neutral (keep the ASIS and pubic symphysis on a vertical plane, with the two top points above the bottom point so they all line up).

Maintaining these points will prevent your pelvis from tilting or twisting—supernecessary if you want to isolate the muscles between the pelvis and thigh.

Once you're set up, lower your pelvis toward the ground by scooting your front foot forward, until you feel your back leg can't extend any farther. The tensile load is now placed on your hip flexors. Repeat other side.

ANTERIOR AND POSTERIOR

In anatomy, *anterior* refers to the front of the body. The opposite of anterior is *posterior*, referring to the back of the human body.

Common cheat: The pelvis, if you're not watching, will tilt anteriorly (imagine pouring a bowl of soup away from you). A tilting pelvis moves the hinge of the lunge away from the hip and to the spine—not where you want it.

Stabilizing the pelvis is enormously important when you're trying to improve the loads created through walking.

• Rainbow Lunge

To create a "rainbow" of hip extension, begin by externally rotating the thigh bone of the back leg all the way (move your ankle toward the opposite side as far as it will go). Do a lunge here with your thigh in this position and again, each time repositioning your ankle (progressing away from the midline) three to four inches at a time. Of course, you can cover more or less distance per adjustment, depending on the time you have. Every degree you rotate your thigh creates a unique load during hip extension.

• Quad Assessment

Most of us have been doing some version of a quad stretch for decades (I remember learning the runner's stretch in third grade), but we pay little attention to the details. Try this out: Standing, see if you can bring the lowest part of your right shin up to your hand *without* arching your back, widening your knees, bringing your knee forward, leaning to one side, or using speed (i.e., kicking your leg up really fast). You probably don't have the length in your quads necessary to reach them without some sort of compensatory mechanism (like deforming some other joint to get there). Try the other side.

• Quad Stretch

Lie on your belly with the ASIS and pubic bone on the floor. Leaving one leg down, bend the other knee, then reach back to grab the ankle without widening the knees or lifting the pubic bone (aka overextending the lumbar spine).

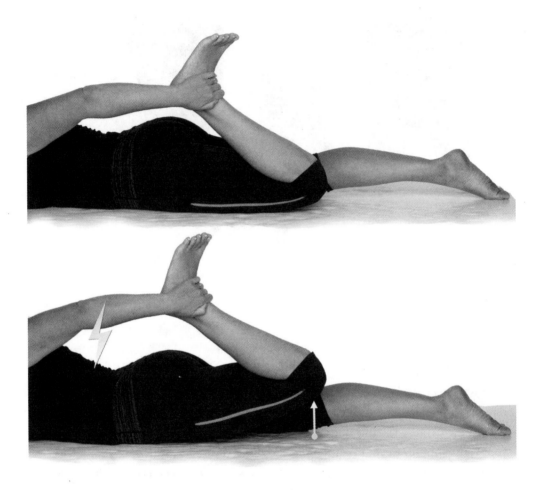

If you can't reach your leg, you can use a strap (or one of the cheating mechanisms listed above), but once you get your foot, make sure to get your pubic bone back down. If your pubic bone is not on the floor, you are bending your knee at the expense of your spine.

Remember, this is about improving the *hip's* range of motion. Stabilize the pelvis!

Want something more? Combine the passive prone hip extension (propping the front of your pelvis on a rolled towel or half foam roller) with this quad stretch—a great way to get all four quadriceps muscles at once.

ANATOMY BIT

The Latin word *vagina* means "a covering, sheath, holder," from the root word *vas*: "cover." In early anatomy, it was used interchangeably with the word *fascia*.

ROTATING THIGHS, QUADS, AND KNEECAPS

All four muscles of the quadriceps pass over the knee joint, become the patellar tendon, and attach to your shin just below your knee. Embedded in the patellar tendon is your patella—also known as your kneecap. Ideally, your kneecap rests in the patellar groove—a special area adapted over millennia to accommodate the patella as your knee flexes and extends. Only, many of us (me included) have spent years walking around with a particular gait pattern created by our super-weak feet and underused hips—a pattern that, over time, creates a gait that rotates the shinbone out relative to the thigh, creating a pull on the kneecap that pulls it out of its groove. This is a lateral (to the side) move that places the patella over a new bony surface without the genetically programmed space that allows for friction-free movement. Now what you have is one bone right on top of another—which means running, walking, or even kneeling hurts.

So. The shinbone needs to rotate back home. The lateral quad and thigh muscles need to reduce their tension patterns and the medial quads—those that pull the patella back to the center of the knee—need to learn to fire again. Sound good?

Although it sounds like something you might order at your local coffee house, the tensor fasciae latae (TFL) is a small muscle up near your hip that tenses the iliotibial band that runs down the outside of your thigh to the knee. Due to extensive sitting, it can adapt to a shortened position (as created by the 90°) and in turn, contribute to a slight but continuous external rotation of the shank.

- **Strap Stretch**

Sitting less is a good way to slowly initiate a change in TFL resting length, but this Strap Stretch can also directly apply a "stretching" force.

Lying on your back with the hamstring of the right leg resting on the ground (if your hamstring doesn't touch, bolster your torso on pillows until it does), lift the left leg and wrap a strap (or belt) around one foot, pulling the foot down to create a 90° ankle angle. (Image left.) Fully extend the knee. Keeping the pelvis level, slowly bring the leg across the body until you feel the load in the lateral hip. (Image right.)

Repeat other side.

- **Rainbow Strap Stretch**

The load to the TFL changes as your thigh rotates. Rotate the thigh all the way out (look at your foot—it should be pointing all the way off to the side) before you bring the leg across. Repeat this exercise, each time rotating your thigh in a bit (as noted by your foot position). Repeat other side.

- **Shank Rotation**

While we typically think of the knee as something that bends or straightens, it also rotates. There isn't a great quantity of motion here, but it is a mobility that helps the knee keep well. As you might find in a moment, in many cases, the shank is excessively rotated outward and held in this malalignment by the tension in the shin muscles. Inward rotation is extremely difficult.

Start seated, with your knees bent. Rotate the entire lower leg toward the midline of the body—which will also turn the foot forward. It helps to manually twist the shin with your hands. Rotate outward (so much easier!) and then work to turn in again.

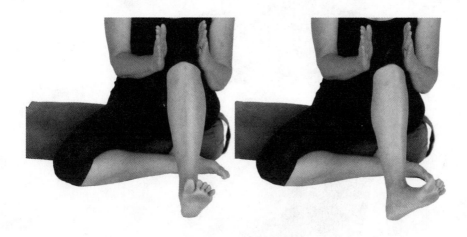

This is one of the few exercises that works best (initially) with a boot that keeps the ankle from moving. Why? Because when we go to rotate the shank it is common to move the foot—creating the illusion of rotation (Look! My foot's moving relative to the ground.) with nothing happening at the knee.

- **Patella Centering**

Lie on your back with one knee bent and the other fully extended on the ground. Before beginning the exercise, the pelvis should be positioned so that the ASIS and pubic symphysis all sit on a horizontal plane. Rotate your femur so that your knee pit (remember your knee pit from chapter 6?) is centered.

Lift your straight leg to the height of the opposite knee without moving the pelvis.

Move slowly, taking your time to lift the leg, and repeat a few times or until fatigue. And did I mention: DON'T MOVE THE PELVIS! If your pelvis tucks back as you lift your leg, you're not working your quads, you're using the psoas. And you just did all those psoas-relaxing exercises, so make sure you're not moving the pelvis and that you are using the quads. Got it? Repeat other side.

FALLING EXPENDS LESS ENERGY

Pushing back isn't the only way we can move forward; we can also lean forward to various degrees and use gravity to create a forward tug on our body—but then walking becomes falling. Fewer muscles used, less metabolic energy (calories) spent, less circulation, less oxygen delivery, and a greater demand on the tissues to dampen the effects of falling. In short, the more we fall, the less we do—which is probably why forward motion by falling is so prevalent.

Leaning forward to initiate movement is beneficial in the short-term, to an extent. (I say *an extent* because falling makes your body more susceptible to injury.) The long-term effects are not at all beneficial. Movement—the translation of the body relative to the ground—by falling occurs without the movement of muscle fibers, without the movement of blood, without the movement of oxygen.

I believe this ability we have to generate lots of movement without utilizing a lot of muscular force is one of the reasons people vary so much in terms of the number of calories they expend when doing an activity. According to your pedometer, you might be walking ten thousand steps every day, only the pedometer doesn't actually monitor how much force you spent walking (read: calories you burned).

Your pedometer only measures how many times your body hits the ground, which makes it entirely possible that *your* ten thousand steps can expend much less (or much more) energy than someone else's. If you are tracking the numbers of calories you are burning and eating and notice that while the numbers should be in your favor but aren't, according to the scale, consider *how* you are moving and not just *how much.*

There is great debate over the generalized calorie-in-vs.-calorie-out theory when it comes to weight loss. Despite this theory seeming like a no-brainer, it isn't. Metabolic science is hugely complex, and there is much left out when we fill the "calorie out" side of this coin with "exercise" and research whole-body metabolic activity with questionnaires, pedometers, and other "easy-to-quantify" but indirect and inaccurate measures of energetic expenditures.

This aspect of metabolic science—that we can generate movement *elastically*, without generating the forces (and burning the calories) we associate with moving—is a complicated research stone still unturned with respect to metabolic diseases like Type 2 diabetes and metabolic syndrome, as well as a major component still missing from the conversation about eating, moving, and body composition.

HIP CHECK: IS THIS THING ON?

Walking is essentially one bout of single-leg balance followed by another and another. When we are on a single leg, this leg—specifically the lateral hip musculature of this leg—must be strong enough to carry the load created by the rest of the body.

The lateral hip muscles are often grouped and referred to as *abductors*, a name that is misleading as it implies that these muscles take the leg bones away from a still body—a use pattern for seated leg abduction or sideways walking with a resistance band. While exercises like these target the lateral hip, they are not specific to how the lateral hip is used during a gait cycle.

While walking, as the single leg receives the weight of the body, the abductor muscles should generate a motion *opposite* to abduction; rather than moving the leg away from the body, they should instead move the body *toward* the planted (or weight-bearing) leg. This pulling of one side of the pelvis downward slightly lifts the other, giving the swing leg room to move forward without needing to excessively bend the knee. Said another way: If you're a big knee-bender when you're walking, you're probably not much of a lateral hip user.

ARM SWING

The work you are doing on your upper body automatically transfers to how your arms participate in your gait cycle. Reciprocal arm swing is the reflexive (you don't have to think about it) backward action of one arm at the same time as the backward action of the opposite leg. This motion is a natural balancing mechanism that reduces the tendency of the spine to rotate, and it saves the spinal muscles from having to overly tense to stabilize the spine with every step.

There are many people, though, who take to pumping their arms up in front of their body for the purpose of working more. This lifting motion is opposite to the naturally backward one. Holding weights can increase the metabolic benefits of a workout, but there is a biological tax immediately collected in the form of compensating loads elsewhere. My advice: Leave the weights at home and increase your workload naturally—by moving more of your body, more often—and let the backs of the upper arms (one of the most underused areas in the body) work in the direction and at the frequency that would keep them in shape naturally!

• Pelvic List

Minus shoes, begin standing with both your feet pointing forward, ankles positioned at pelvic width—the distance between your right and left ASIS. Your weight should be shifted back over your heels and your legs should be vertical.

Place your right hand on your right hip and try to bring the right side of the pelvis down to the floor. This downward action will result in your left foot clearing the ground. Check to make sure you're not bending either knee, and relax your knee-caps! (Image at right.)

Is the standing ankle wobbling, or the floating foot touching down periodically? Does the pelvis thrust or tuck or knees slightly bend? Do the arms become involved to regain balance? These are all signs that you don't regularly recruit your abductors when walking. To remedy wobbly ankles and a lack of balance, do this exercise more often and hold for longer periods of time.

Cheat alert: It is possible to attain the "look" of this exercise by hiking up the floating side of the pelvis with the muscles of the lower back. Same move, different muscle groups. Placing your hand on the working (standing) hip can help remind you which muscles should be creating the movement.

LIST

An inclination to one side, like a wave tilting a ship to either port or starboard. When you're standing, a pelvic list would result in one side of the pelvis being higher than the other.

FOOT LEAVES
THE GROUND

- Next-Level List

Stand on a phone book or yoga block with the other leg floating above the ground. You'll follow the same listing procedure, but the extra height increases this exercise's range of motion, which better prepares these muscles for going up and down hills.

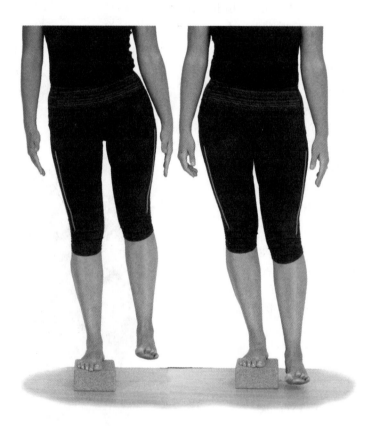

HIP STRENGTH IS NOT ARBITRARY

Ideally, lateral hip musculature should have the strength and endurance to repeatedly elevate the pelvis (under the weight of your torso and other leg). These small hip muscles have to be strong enough to hold essentially the entire body minus the standing leg. We're talking SERIOUS strength.

But what most people do to exercise the lateral hip are abduction exercises—think Jane Fonda leg lifts—where these muscles are trained to hold the mass of one leg. If you quickly do the math, you'll find that while abduction exercises strengthen abductors enough to lift the weight of the leg (plus whatever extra resistance a tube or machine offers), that weight is much less than the weight of the entire body (minus the leg still on the ground) under the gravitational loads (or Gs) created while walking—almost double the body's weight while standing. Our legs are totally unprepared for the demands placed upon them for walking.

SWING LEG

Swing is the phase of walking in which one leg is in the air for limb advancement. The swing leg is the non-weight-bearing leg when you're taking a step.

Walking, in order to be an efficient use of energy and of the tissues of the body, should be as streamlined as possible. There are many things our bodies do when they're supposed to be moving forward—they lurch side to side, they fall a little and need to be "caught" with each step, instead of front and back our

ANATOMY BIT

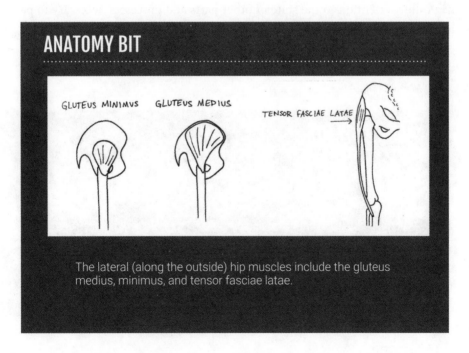

GLUTEUS MINIMUS   GLUTEUS MEDIUS

TENSOR FASCIAE LATAE

The lateral (along the outside) hip muscles include the gluteus medius, minimus, and tensor fasciae latae.

arms swing right to left or maybe don't move at all. Each of these motions moves the body's center of mass unnecessarily. Whether it's side-to-side or up-and-down, unnecessary motions consume energy and break down joint tissues without contributing to the body's desired net displacement, which is forward.

WALKING IS NOT OPTIONAL

Everything you have done so far—adjusting your head, strengthening and mobilizing your arms, releasing your psoas, and preparing your feet and ankles—will be directly impacting the way you walk. Instead of thinking about how to walk "right," it is better to release your immobilities and constantly expose those new ranges of motion to the activity of walking.

I am always surprised when people say they find walking boring. Walking defines us as a species. It is not a luxury. Not a bonus. It is *not optional*. Walking is a biological imperative, like eating and having sex. Which is why we should, as a species, see the inability to walk without pain for what it is—a huge, red, waving flag calling attention to the state of other parts and processes necessary to perpetuate our humanness.

Which brings us to the pelvis.

ANATOMY BIT

Many people are able to ride a bike without it falling over, yet only a very few can sit on an unmoving bike and say the same. Why? Continuous forward movement can mask myriad balance deficiencies. Just as the cyclist makes superfast corrections to a thousand invisible almost-falls, someone walking with a body weakened by modern living makes corrections to one fall after another. These falls occur too quickly to see, if you're trying to evaluate them by observing someone's walk. The stance phase of a gait cycle can be isolated, however, as it is essentially the same as standing on one leg. When you stand on one leg—no bending, no arms out to the side—how do you do?

NOT YOUR GREAT-GREAT-GREAT-GREAT-GREAT-GREAT-GREAT-GREAT-GREAT-GREAT-GRANDPARENTS' PELVIS
CHAPTER 10

The very act of "not squatting" may be as important in terms of creating a biomechanical environment as squatting itself. **R. S. KIDD AND C. E. OXNARD**

This is the chapter about the pelvis. You might have noticed that I have not dedicated an entire chapter to a single body part until now. If I had, this would have been a very long book. The pelvis, however, is different. And in all fairness, this chapter is not about the pelvis *as a part*, but how the pelvis integrates—*should* be integrating—into almost everything your body does.

If we look at humans from an evolutionary point of view—heck, even if we look at humans from a *non*evolutionary point of view—procreation is, literally, our population's defining act. Even if you don't have any children, someone had you; each of us is separated by a maximum of one degree from this reproductive phenomenon—a phenomenon that depends entirely on the pelvis.

The pelvis is the skeletal environment for sexual pleasure and performance, egg production, sperm production (read: fertility), menstruation, pregnancy, and delivery. I am most passionate about the pelvis and its issues because *the success of the pelvic area is the success of our species.* The pelvis, for both man and woman, is foundational to moving DNA on to the next generation.

But setting aside the obvious pelvis-users like sex and birth, what about other important functions? The pelvis is the final skeletal stop for everything you eat and drink. Sitting, standing, and getting down and up from the floor are all motions that pass through the pelvis. Your pelvis sets the base for your spinal column, holds the weight of your torso, and connects your upper body to your lower.

Delving deeper, the muscular floor of the pelvis (the pelvic floor) is every torso's force-generating basement. The pelvic floor is a group of muscles responding to the weights and forces created by the organs and inner workings of the thoracic, abdominal, and pelvic cavities. In the best of scenarios, the pelvic floor responds to keep our organs supported—preventing prolapse (the falling out) of organs in women and the falling down (as in, onto the prostate) of organs in men—while allowing the easy rotation of our hip joints and supplying blood flow to all the fun bits. I'm not really a fan of creating a Who's Who list amongst body parts, but if I were to play favorites, I'd probably go with the pelvis.

In addition to writing an entire chapter about the pelvis, I've also saved this chapter for the end. Not because I wanted a surprise ending, but because I do believe that the pelvis, more than any other part of the body, is struggling out of nature in the same way an orca fin struggles in captivity.

THE INTEGRAL PELVIS

Housing the center of your mass when standing upright, your pelvis is affected by everything you do. The shoes you wear, how you sit, *how much* you sit, how you walk, *how much* you walk, the stress you are under, your beliefs and practices with respect to posture, and all the ways you move—or don't—have molded your pelvis into the shape it is right now. The good news is, everything you have learned in this book so far—making over your stance, gait, upper-body mobility, etc.—also impacts the pelvis.

We aren't used to working on our bodies this way, I know. We tend to approach pelvic problems by focusing on the muscles in the pelvis and tend to ignore (or be ignorant of) the impact of forces outside of "the pelvic muscles" on pelvic goings-on. You might do Kegels daily, not realizing that the very way you breathe—shallow or erratic breathing, stemming from stress- or kyphosis-induced changes to lung inflation—is constantly placing excessive pressure on the pelvic floor. Or maybe you'll spend an hour a day working on a pelvic and core strengthening routine, only to stroll out the door in heeled shoes at the end of your workout, oblivious to the sarcomeres weakened by how your rib cage, pelvis, and trunk muscles have to adjust to compensate for your new ankle geometry.

The pelvis, like all your other "pieces," is not a mysteriously broken part living in an otherwise healthy machine. The pelvis is simply responding to the whole-body communication you create through your daily behavior. For this reason, you will not find a list of "pelvic strengthening" exercises in this chapter. What you will find here is an explanation of how whole-body movement issues—specifically the lack of walking, the lack of hip extension while walking, and the lost practice of squatting—affect the pelvis, and how this area is best served by restoring the natural loads to the whole body rather than spot-treating symptoms of weakness.

YOUR PELVIS, WALKING

Most therapies agree that the bulk of pelvic issues arise when the position of the pelvis is faulty. Thus a common therapy is to return the pelvis to neutral (as already defined in chapter 6). But the static position of "neutral" isn't a magical "now I can work" position. A neutral pelvis sets the stage for the greatest amount of loaded hip extension.

Walking is kind of like rowing a boat. The oar submerges, pushes back against the water, which propels the boat forward. The oar comes up and forward to repeat the cycle. When it comes to walking, you "row" your body forward using a single "oar" (leg) at a time.

In order to do this, the entire weight of the body must be held by the lateral hip muscles of the one leg that is pushing back against the ground. This allows the oar on the other side of the body to get into place to continue the forward motion once the full backward motion has been completed. This phase of carrying the body's weight while pushing back is called loaded hip extension. But our so-called "walking" muscles aren't only creating the walking motion.

These muscles, as they walk us forward, actively stabilize the sacrum in the pelvis, and provide a counter-resistance for the pelvic floor (which experiences greater loads when you're walking than when you are still), optimizing the leverage of your organ support. Your hip muscles do this—hold the weight of your body in the lateral hip, extend the thigh bone, stabilize the pelvis and sacrum, and support the function of the pelvic floor—*all at the same time*. It's a very specific and organic process that requires less time and energy than if you needed separate "strengthening" activities for each of these parts individually.

In the same way the byproduct of natural swimming is an orca fin's uprightness, pelvic floor function and sacral and pelvic stabilization are the *natural byproduct* of a lifetime of walking (while not wearing heeled shoes, or even shoes at all!), squatting, and not using chairs.

Pelvic issues, for both men and women, are hugely prevalent, yet the mechanical environment of the pelvis as it relates to the whole body is rarely considered, and the "pelvis out of water" (I mean, "out of nature") is only recently being examined in research literature with respect to the elimination process and pelvic floor issues that arise from not squatting to evacuate. I expect there will be much more research in this area in the next decade.

WHERE HAVE ALL THE BUTTS GONE?

Just as the response to muscle use (think: bicep curl) is an increase in muscle mass and strength, the response to hip extension is more mass on the backside. You might not spend as much time checking out everyone's butts as I, but you should. In fact, put this book down and head somewhere where there are a lot of people (like an airport); you will see that there is a definite case of missing muscle mass that needs to be solved.

But then again, if you put everyone in the airport (you're now reading this at the airport, right?) through the battery of hip-extension tests and exercises contained in this book, you'd find that most no longer have hip extension available. The seemingly bottomless (!) pool of glute-less people matches up exactly to the number of those without hip extension.

Of course, even with ideal hip extension, not everyone would have bulbous glutes, so don't go around judging people by how their backside measures up to someone else's. We all have a different absolute mass that would develop,

ARE ALL BUTTS EQUAL?

So, we don't have butts. Which isn't to say that we can't go into a gym and carefully isolate these muscles by mindfully watching our form and working muscles to keep our tension from pulling our pelvis out of alignment, so that we can carefully body-build and develop one. But the natural frequency and amount that the posterior leg muscles should be working would not occur. If you figure your butt muscles should be working with each step—and you should be taking enough steps to cover three to five miles per day—you can see how it is not likely that the three sets of a hundred reps of whatever exercises you are doing for your hip area match the natural loads necessary for the lumbopelvic area to function optimally, to function biologically.

Many have crafted a backside by working diligently on maintaining a neutral pelvis by forcing its position despite their limitation in hip extension. In this case of Tension vs. Tension, not only are the hip flexors still tight on the front of the body, the glutes are conditioned to the abnormal amounts of tension necessary to "balance" the problem. In the end (get it?) your new butt isn't the butt (on a cellular level) that you were after, although it might fill your jeans in the same way. This method of "body balancing" isn't a viable long-term solution, as now there are two muscle groups experiencing abnormally high loads. Having too much tension on both sides of a joint can lead to an entirely different set of problems.

depending on frame size and body weight were we to walk without the adaptations to our lifetime of chair use. That said, everyone should have their fair share of glute mass. It's a better idea to measure gait patterns than glute size, but until you've had training in how to quantify walking, it's easier to see a lack of butt muscle than it is to spot a gait pattern lacking hip extension.

PELVIC ANATOMY IS WHOLE-BODY ANATOMY

There are over thirty muscles that attach to the pelvis, sacrum, or hip bones, and every one of them affects the biomechanical environment of the pelvis. The good news is, you've already been working to undo many of these inappropriate tensions and weaknesses by adjusting your gait, sitting practices, footwear, and upper-body mechanics! You've already been working on improving your pelvic health, *and you haven't even done the exercises in the pelvic chapter yet!*

So, in addition to all the other things you are doing, do these exercises to target

· · · · · · · · · · · · · · · · · · · ·

PRONE

To lie, face down. Opposite of
supine.

· · · · · · · · · · · · · · · · · · · ·

any muscle preventing the pelvis from easily accessing
its neutral position during the gait cycle, as well as
muscles that may be overly binding the sacrum to the
hip.

- **Prone Inner Thigh**

Lie face down with your belly on the floor. Slowly scoot one leg out to the side
without bending the knees and with minimal lifting of one side of your pelvis
away from the floor. Inner-thigh tension will rotate your thigh down into the
floor. As you advance, see if you can externally rotate (rolling your thigh bone so
the toes on that foot point more toward the ceiling and less toward the floor).
Relax your head and neck on your hands.

- **Legs on the Wall**

First, find a wall. Some people have avoided this exercise, saying, "I don't have
any free wall space." If your home is especially small (I had a client who lived
on a boat) or extremely cluttered, doing this on your bed with your legs on the

headboard could also work, although the softness of the bed skews the loads a bit, so it's not exactly the same stretch. You might want to pack up some of your knick-knacks until you've improved the health of your hips, dig?

Sit sideways to the wall and then rotate your body to get your legs straight up the wall. If your hamstrings are especially tight, your pelvis will tilt backwards,

forcing your waistband to the ground. In this case, back your body away from the wall until your pelvic markers (ASIS and pubic symphysis) line up in a horizontal plane. This will put a bit of space under the top of your sacrum.

Keeping your legs straight, relax your legs away from each other until you feel a "stretch" in the inner thigh. This can be an intense exercise, so rest by bringing your legs together when you need to and resume when you are able.

· **Number 4 Stretch**

You've probably been introduced to a version of the Number 4 Stretch at some point in your life, but here are two different versions to try:

On the ground (easier version): Lie on your back and bend both knees, keeping the soles of your feet on the floor. Cross one ankle over so that it rests on the lower thigh of the opposite leg. Make sure you aren't tucking the pelvis. If this is too difficult, you can slide the foot (still on the ground) away from you. If it is too easy, you can pull the bottom leg off the floor and toward the chest. Make sure you aren't tucking, hiking, or twisting the pelvis in any way. This motion should be created at the hip joint.

In a chair: Sit with your bottom scooted to the front of a chair. Cross your right ankle over the left knee, careful not to twist, tuck, or hike the pelvis in any way.

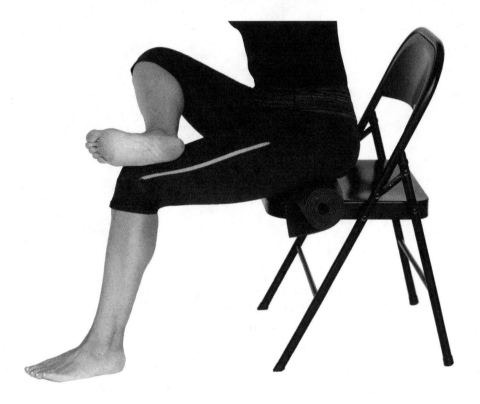

If your knee takes the brunt of this stretch, then your muscle tension is too great for this exercise; go back to doing it on the floor. If you can feel it in your thigh (anywhere—the outside, the groin, the hamstring), then tilt your pelvis anteriorly (to the front) a bit to increase the stretch.

- Reclined Sole-to-Sole Sit

This and the next exercise are an extension of the stretches found in chapter 6. Instead of sitting up with the soles of your feet touching, lie back on the ground (bolstering your head and shoulders in the same way used for minimizing rib thrust in the upper-body exercises). Lying back not only takes your torso back, it also takes your pelvis, introducing a different load to the connective tissue, muscle, and bone in the knees, hips, pelvis, and abdomen.

- Reclined Cross-legged Position

If your legs are already stretched to the max in regular cross-legged sitting, then reclining your pelvis back to the ground will probably have to be done in steps. Prop your torso up to the angle where you first begin to feel the limitation of your body tension. This is where you can start, lowering your torso over time to match your new mobilities.

NOT SQUATTING

I lecture, often, on the mechanics of the pelvis and all of the muscles that impact the inner workings of the trunk. My lectures always reference the impact that *not squatting* has on these systems. After one lecture, a lovely physical therapist came up to me and asked about the contraindications of squatting. She had learned in school that women with prolapsing organs should avoid squatting as the strain could worsen the issue. I wholeheartedly agreed with her. And I also threw in that squatting is listed as "contraindicated" for those with bad knees, bad backs, and bad hips, as the loads could worsen those issues as well. "But," I said, "I also believe that squatting is a non-negotiable ingredient to improving issues with the gut, pelvis, hips, and knees." The problem, you see, is not the squat but that we haven't squatted—for the bulk of our lives.

But before I go any further, I must explain that the word *squat* does not imply a single exercise, but rather a *category* of moves typically including some amount of knee and hip flexion.

In fitness, a squat is typically an exercise done repetitively (as in, three sets of twelve) with a form (knees over ankles!) selected to minimize inappropriate forces to the knees and maximize the use of the glutes. The farther forward your knee goes past your ankle, the less glutes you can use and the more the quads have to contract, which, as you learned earlier, pulls the patella backward into its groove. Except, in our modern bodies, the patella doesn't sit over the groove but slightly to the side—which is one of the reasons squats are typically contraindicated for those with hip or knee issues. When you add a bunch of repetitions or place extra weight on this move through other fitness accessories like hand weights or weighted bars on the back, form becomes even more important, as this move isn't a naturally occurring phenomenon.

The "deep" squat is usually one that utilizes the full range of hip and knee flexion. This could also be called the "hunter-gatherer squat" or the "bathrooming squat," although I'd like to clarify that hunter-gatherers use all sorts of different squats (so the squat used to bathroom in places without toilets is not automatically the same squat geometrically as the one used to dig tubers).

Other squats could include yoga poses where you maintain a chair-like position without a chair, or dropping all the way down into a squat with the knees very wide and toes turned out, hovering over the public toilet seat, or any fancy (one-legged!) variation used to build lower body strength or skill. It's important to remember that while each of these fall into the same "squat" category, they are all different actions that use the body differently and result in a different tissue adaptation.

The progression to the squat, in this book, is referring to the deep or "traditional" squat—heels down and maximal hip and knee flexion without excessive distortion (wide turnout, for example) of the feet, knees, or pelvis.

I've heard many professionals say that no one should be doing a deep squat as it "wrecks your knees." I disagree. Humans all over the world use deep squats multiple times a day, for extended periods of time, throughout a lifetime. But on the other hand, I also know that deep squats, when done with bodies adapted to not squatting, create an entirely different load profile on the cellular level than a squat done by a body with access to full ranges of motion. In most cases, squats tax the elastic properties of our muscle.

Just as we can force knee flexion beyond our current range of motion by pushing muscle into its elastic properties, we can use gravity to force our joints to bend more than our muscles can relax to accommodate. Or, we can easily get down into a range of motion by falling, but when it comes time to get up we have to heave-ho. Both of these scenarios can make a deep squat more damaging than

ANATOMY BIT

Heterotopic ossification is the process of developing bone at an abnormal anatomical site, as in the case of a bone spur. The cavalryman's osteoma (also known as "rider's bone") is another example of a heterotopic ossification—in this case a slow-growing tumor of bone in some of the inner thigh muscles in response to unnaturally frequent pressures created by straddling a horse. This type of bony response to behavior is different from bony features that are potentially phylogenetic, like squatting facets.

Unlike naturally occurring bony formations, heterotopic ossifications are often considered pathologies, and after time are associated with symptoms (like pain), as an increase in mass on one area will trickle down to the distortion of surrounding tissues.

Facets, on the other hand, are reflexive patterns in bone that result from the movement habits of our ancestors, and they are there to *allow* for aligned movement—that is, movement that doesn't create unnecessary tissue breakdown or damage to other parts.

beneficial. So while an orthopedic professional might say, "Deep squats are not a healthy exercise and can lead to damaged tissue," it would be more accurate to say that squatting is an entirely natural human movement (as are walking, pooping, and chewing your food), but your body, as it is right now, is not able perform a deep squat in its most natural sense without doing harm, and what wrecks your knees is you taking your body in its current form and loading it beyond your tissue's current capacity, which is why we say not to do that.

See the difference?

While I do think that, for most of you, there can and will be plenty of squatting in your future, there is a very strong chance (such a strong chance I would bet a ton of money on it) that you currently do not have the equipment necessary to squat a lot without some sort of biological tax.

. . . .

PHYLOGENETIC

Based on natural evolutionary relationships.

. . . .

Probably you've got some of what you need, like your leg bones, ankles, knees, and hips, but there are smaller pieces of your squat anatomy missing. Parts you once had and eradicated through a lack of use (like squatting facets, which I address in the next section). There are also essential muscle masses (in the legs, butt, and upper body) that would change the weight and balance of your body, creating a different joint orientation to your squat which, in turn, affects the muscles you use to get down and back up again.

Your muscles are tighter (read: literally shorter or longer than necessary), your joint motions are more limited, and your masses are higher or lower than they would have been in a natural setting. For this reason, the rest of this chapter is the step-by-step manual to aligning the cells of your body so that you are able to squat, naturally. Are squats contraindicated? Sure. If you've got artificial hips and knees, fused parts, or anything else that has permanently altered your joints' range of motion, there is no way these can be undone—no way for the limitations set by metal to be modified. Still, this chapter breaks down the squat enough that *everyone* can get a little closer to a squat and reap its health benefits.

THE BODY DOES WHAT IT IS TOLD

FACET

In bone, a small, smooth-surfaced process for articulation.

From birth (*in utero*, actually) your bones included facets that allowed for the joint articulations necessary for squatting. This is one of the reasons squats come so naturally to children—there is nothing structural limiting this type of joint action. Bones, growing under the conditions of squatting, maintain their squat-friendly shape as they become larger and more dense. Growing up squatting means a child-then-preteen-then-post-adolescent-then-adult bone continues to contain these "squatter's grooves," these grooves being mechanotransduced by the act of squatting. Without the specific push and pull on these tissues, the communication created via mechanotransduction amounts to "no groove necessary."

As we explored earlier in the book, our chronic position and lack of movement have left us with an altered "falling" gait pattern that fails to include hip extension. Falling in lieu of walking, in turn, has resulted in not only a lack of total muscle mass, but also a different distribution of our existing lean mass, especially about the hips and thighs. Our missing mass is a problem not only

CONTINUUM

A continuous sequence in which adjacent elements are not perceptibly different from each other, although the extremes are quite distinct.

when it comes to balancing a squat; there's also less muscle there to assist in force generation when it comes time to lower down or come up. Your body shape as it is now—under-muscled and without full and smooth articulation of your joints—means the squat that you can do (if you can, in fact, do one) is not the same "nutrient profile" created by the squat you would have done had you always done them.

To go from being a non-squatter to being a squatter creates unnaturally high loads (due to a lack of facets, muscles adapted to shorter or longer lengths, and inhibited joint range of motion) on the involved structures like your bones and joints. Maintaining or rising from a squat—especially when you are missing the muscles that get you back up—tends to create tremendous downward loads on the (gulp) gut and pelvic floor. And we're doing these for our health, right?

This book is about natural loads, not exercises. If we are to create a tissue environment that is most similar to what it would have been had we been raised in nature, and most similar to the state in which the body's functions flourish, we have to put squats on a continuum—where we can gradually work to communicate our intentions and reshape, over time, our tissue so that our body-autobiography once again says *squatter*.

It is most helpful to think of a squat as a journey and not a destination, both in the grand sense (i.e., you're probably going to be doing a modified squat for some time) and in the literal interpretation of the squat. You could define a squat as a position in which one's knees and hips are fully flexed. But if we say a squat is an essential human motion and define it as a fixed position, we again reinforce that there is an inherent benefit to a position. It's like reducing a human food (say, an apple), to a single nutrient (such as fructose). How does fructose behave outside the context of an apple? Reducing, in our minds, the squat to a single joint configuration takes the squat out of the movement context and makes it more challenging to see the full spectrum of its benefits.

If we expand our definition of a squat to include the "getting down" and "getting back up" phase, we can see that the body is used differently for controlling our downward motion, hanging out in knee and hip flexion, and coming back up again. Each of these phases creates different and essential loads. A squat is much more than a position; a squat is an experience.

PREPPING YOUR SQUAT

Skills necessary for squatting can be broken down into two categories: ranges of motion and strength throughout those ranges of motion. Many people have no trouble with joint mobility, yet they lack the muscle mass (or coordination) to control their motion as they lower down. And, when it is time to come up, people lacking strength use a lot of momentum (and typically a lot of bearing down).

Some have the strength to squat, but not the mobility. They can control their motion toward the floor (and not crash down) and easily come up without a big heave-ho. But without full joint ranges of motion—of the ankles, knees, hips, pelvis, and spine—these people might be using their existing patterns of moving and strength and using a geometry that prevents the underused muscles—especially the glutes—from participating. For these reasons, we begin with unweighted postures and positions that introduce joint ranges of motion without high loads that can create damage.

· **On Your Back Squat**

Lie on your back, then bring your knees in toward your chest.

Easy, right? Now, check to see how much your pelvis tucked when you brought your knees in. Do this again, only don't let the pelvis leave the ground. Only pull your knees in to the point that your pelvis wants to come with. Squatting, just like walking, will utilize the mobilities of all your joints, but when restoring natural squat function, you have to break up some of the sticky relationships that are limiting the action of some of your parts—in this case, the pelvis that moves to compensate for hips that don't.

Dear people with joint replacements and knees and hips (or feet or a spine) that feel so far gone that you can't do anything: Even if you only do this—lie on your back, in bed even, and bring your knees and hips to your chest, and even if you only do it one leg at a time—your body will become better. You don't have to become Super Squatty McSquatter to improve. You begin improving *the moment* you change loads!

SQUAT-TO-POT

Squatting to go to the bathroom is so crucial to pelvic health, some pelvic restoration protocols now include it in their therapeutic protocol for those with digestive and pelvic floor issues. May I suggest that we include it in our preventive model as well?

Despite the squat being much more than a position, a great deal of the benefit to squatting is, in fact, the position. If we take away the muscle-firing necessary to squat, there are still benefits when it comes to elimination. For example, the orientation of the anorectal angle changes the pressures necessary to eliminate. If your tubes aren't lined up to take advantage of gravity, some other force needs to be created to get your stuff out. As we age, that pressure necessary to eliminate becomes a liability, as this straining is a causative factor in "death by toileting," also known as cardiovascular events (heart attacks, strokes, aneurysms) triggered by the mechanical environment created by the Valsalva maneuver.

Even before you have the strength and mobility to do a squat, you can enjoy better-aligned toileting by putting your feet up on something to increase the bend in your knees and hips while you poop. There are specific products (low-profile platforms) that pop on to the front of your toilet—no tools necessary—or you can just flip over a couple of small trash cans or haul a couple of stacks of books to keep on either side of the commode. Either way, this is a huge body-saver that takes no extra time out of your day at all. I mean, you're already sitting there on the toilet, right? Why not get some squat-prep stuff done in your regular life?

WORKING TOWARD YOUR SQUAT

Again, all the other things that you are working on—calf stretching, lunges, working on your shank rotation—play a role in the whole-body ranges of motion necessary to squat. This chapter's exercises are designed to create more specific loads necessary to reshape you into a squatter, but are meant to be done in the context of all the other exercises detailed in this book. This series of squat-prep exercises can be used as a twenty-minute warm-up for a squat session, but are probably even more appropriately used as a six-month (or three-year) warm-up to doing your first squat.

- **Prone Squat**

Beginning on your hands and knees, ease back until your hips rest on your feet (or as far back as you can go) and relax there. If you are very new to this motion and it is causing your body to tense or hurt, reduce the load in the following ways: do it on a bed or other soft or squishy surface, place a pillow behind your knees (so they aren't forced into flexion), or place a pillow on top of your feet (under your hips) so that gravity doesn't force you beyond your current ranges of motion.

- **Prone Squat, Untucked**

Using a mirror to view yourself from the side, start on your hands and knees. Lower your bottom toward your heels, allowing your knees and hips to flex until you see your pelvis beginning to tuck. Without overworking the spine, prevent your pelvis from tucking while maintaining the position. Your pelvis starts tucking because the muscles of the hip can no longer lengthen and shorten to accommodate a full squat position; they're adapted to huge frequencies of still-ness. The point where your pelvis starts to tuck marks the boundary of your hip's current range of motion.

- **Prone Squat With Dorsiflexion**

Again, from your hands and knees, sit back, only with your feet tucked to decrease the angle at the ankle. You can do this with your pelvis both tucked (more knee flexion) and untucked (less knee flexion).

- **Calf Stretch Soleus**

The Calf Stretch you've been working on targets the gastrocnemius muscle of the calf group by keeping the knee from bending. But there is another muscle in the calf group, called the soleus, that is better stretched with a bent knee. Your soleus has likely adapted to positive-heeled shoes, and this is one of the main reasons keeping your heels down in a squat is so challenging. To increase the ankle's range of motion necessary for a squat, try this.

Standing on the half foam roller with the ball of one foot up on the dome and its heel on the ground, bend the same leg's knee—pushing it slightly forward—as you press that same heel toward the ground.

THE NEW YOU REQUIRES A NEW SQUAT

Before we go deep into the Deep Squat, I want to stress once again that all the other exercises in this book are also part of your "squat prep" plan, although more indirectly. Once you have greater mobility and strength in all your parts, the action phase of a squat can be done with more support and less "tax."

If I had you get down into a squat right now, the foot, ankle, and shin positions that come easiest to you say a lot about how you have moved your body up to this point. (If you're okay with squatting, you can go ahead and try this right now, in front of a mirror. Do what you need to do to squat and take stock of your foot, ankle, knee, and hip positions, looking at both the side and front view.)

How much do your feet turn out relative to your shin? More than when you are standing? Is all your weight on your toes or can you lift the front of your feet? Do your heels even touch the ground?

Did your shins rotate inward, collapsing the ankle joint and foot? Is the instep of your foot pressing more into the ground than the outer edge?

Do your knees have to widen? Come together? How much? More on one side than the other?

Is your pelvis forced to tuck under, and is your lower spine so rounded that your butt is completely underneath you?

The turnout of your feet, the collapse of your ankles, the amount your pelvis tucks, and the angle at which your shins project all communicate, biomechanically, the state of your tissues below your skin. While this way of squatting comes naturally to you right now, this position isn't really an indication of You, in nature. This is why I include parameters for foot, ankle, shin, and pelvis positions when squatting. If you continue to squat following the boundaries set by how you have moved in the past, your body will be continuously shaped by what you have done before. The squat we are after is one that slowly coaxes out a version of yourself not tainted by modern living.

The first of these parameters is foot position. Many people have an abnormal amount of turnout in both the foot and shin that has been cemented by a gait pattern that uses no foot or hip musculature. It takes years of mindful moving and a complete habit overhaul to correct foot turnout *while* walking, but it's much easier to make incremental adjustments to your feet and knees when you are focused on a series of corrective exercises. Alignment points not only keep you working the muscles these exercises are designed to target, but also they prevent the existing adaptations to your body—the twists and turns in your bones, the lengths of your ill-moving muscle, and the mental movement habits that come

ANATOMY BIT

The vertical shin is a simple indicator of which muscles are being used on the way up and down as you squat—think of it as an inexpensive biofeedback tool. The more vertical the shin, the more the glutes can drive the motion. The more your knees move in front of your ankle, the more quadriceps are handling the work.

easily to you—from affecting the new shape you are working towards.

When you squat, take note of your most-frequented foot position and turn it inward (to whatever degree is tolerable) over time, until the foot more or less lines up with the angle of the thigh.

And speaking of the thigh, note any tendency to widen your knees excessively. Ideally, your knees should be just slightly wider than your pelvis. Going much wider than this will bypass the tensions in your hip and thigh that limit your ability to flex your hips and knees in one plane, by taking everything to an entirely different plane of action. If you are a wide squatter, work, over time, to narrow up your stance.

READY, SET, (ALMOST) SQUAT!

This next series of squat exercises is more advanced, in that they are full-weight-bearing. Gravity, in this case, can push your body beyond its limits, and this effect is not as easy to control, so before trying these exercises, make sure you read through the entire section to get a sense of how to assess where your strengths currently lie (and where they do not) as well as a sensible progression.

• The Chair Squat

I'm thinking that unless you have oodles of free time, you're going to be doing your "squat program" once a day at most. So let me start this next session by highlighting that you're already getting up and down out of chairs (and eventually off the floor, right?) a lot throughout the day. How about turning this time into squat work?

When we mindlessly get up and down out of a chair, our current tissue adaptations result in us using our arms to push or pull against something to assist, jerking forward to create momentum, or shooting our knees way out in front of our ankles to let the quads do the work. Instead of continuing to get up and down in ways that allow your muscles to escape use, take advantage of this time—already built into your day—to build your butt!

Begin sitting near the front edge of a chair, with your shins vertical, feet flat on the ground. Untuck your pelvis and lean forward (not quickly). Without letting your knees (and shins) slide forward, see if your butt is strong enough to get you out of the chair. You can also try to lower yourself down (no crashing to your seat) to practice the control you'll eventually need for a full squat.

If you have difficulty getting up and down out of chairs using your glutes, imagine how much you'd have to strain (bear down) or call on the front of your body to get up and down from a full squat! Spend a few months doing this exercise as often as you can, trying it on seats of different heights: Once getting up out of your kitchen chair is easy, find a crate and begin introducing your muscles to a new range of motion. Eventually you'll have the strength to do the full range of motion, but the great news is, even if you never progress beyond chair squats, your body is already more aligned with the natural version of yourself.

Your muscle strength, joint range of motion, circulation, and tissue adaptations have been in progress since you started. I've had many clients never move on due to joint replacements or simply a desire to work where they were comfortable and yet still experience a huge improvement in health. There is no doing it right, but there is always doing it better!

When you're ready to move on to full squats, whenever (if ever) that is, the following is a series to assist you in getting most of your parts into the Deep Squat while using less of your muscles' elastic properties.

• The "Do It However You Can" Squat

This first deep squat doesn't really have a recommended set of alignment points beyond the feet position and knee width. Without worrying about keeping your feet down or pelvis in a particular place, just play with your balance and control in this position.

- **Full Squatting with Bolstered Feet**

The adaptations of the calf muscles in response to shoes can limit the articulations of the rest of the body during a squat. To eliminate the plantarflexing
(toe-pointing) forces that, while squatting, tend to knock you backward, put
your heels up on a phone book or foam roller. The higher your heels are elevated,
the less your squat is affected by the shortness of your Achilles' tendon. You can
work on your feet another time. For now, just enjoy.

- **Full Squatting, Untucking, and Backing Up**

Remember those missing glutes? Without glutes, a squat has to be driven
mostly by the quads. To slowly challenge your backside, you must think "untuck
the pelvis and back up the hips" as you come up and down. There will be a point

in squatting (both up and down) in which you can no longer maintain these motions without falling backward, but if you are always trying right up to that point, you will, over time, strengthen and increase the mass of the butt muscles and develop a more well-rounded (butt pun!) way of using your hips.

• Assisted Squatting

Once you've made progress on your feet and glutes, you can fill in some of the strength gaps by using light assistance. When you first learned to ride a bike, an adult's hand on the back of your seat gave you just a little bit of the support needed to solidify a new motor skill. In the same way, holding (lightly) on to something as you shift back, even more, onto your heels while untucking your pelvis can help you recruit muscles necessary for stability in this position.

The motion of backing up and untucking doesn't need to be visible to be "feel-able" to these muscles. Changes in strengths occur even when change is incremental. You might not look untucked and backed up, but as long as you are a hair more untucked/backed up than you were before, the loads will reshape you to make this motion even easier the next time you squat. Repeat, repeat, repeat.

Here I've used a chair, but you can also do this by holding on to a pole, door-knob, or my favorite—a TRX strap!

Because we are all unique in size, shape, and starting point, there is not an absolute form that could be called "the best" squat; however, there are relative variables that make one squat better than another. For a more even use of parts (read: load distribution), squats should, ideally, have a basic symmetry between feet, ankles, knees, and hips. If you notice that in order to go deep you have to create a radical distortion to one side but not the other, it is likely that your habits have created asymmetrical sticky spots. Deal with them for better squatting (amongst other reasons).

Straining a lot during squatting—on the way up, down, or while maintaining position—is a flag that your squat is exceeding your capability. Yes, as you transition your body small strains will help you adapt, but enduring large strains or experiencing pain are indications that you have taken a step too large for your physiology at this time. Take a step back (not literally) and squat at a more appropriate level.

Squats that are "whole body," meaning the work necessary to lower and lift your body is shared well across your joints and muscles, occur when your muscle lengths (read: number of sarcomeres in a chain) allow full use and force production of both sides of a muscle. The amount your pelvis untucks and the ability you have to move your hips back toward your heels is a good way to measure your progress toward greater body use.

Once you have practiced, practiced, and practiced, your best reshaping squat is the one where you are farther back in your heels, use less momentum, and are more symmetrical than you were before.

The more you squat and the more you mind how you get down and back up again, the stronger you will become and the less momentum you will need. This takes time, but it's actually more accurate to say squat-competence takes frequency: many squats, spread throughout every day, as you would find them while living in a natural setting.

THE SQUAT, AS A LIFESTYLE

Taking a snapshot of a squat for the purpose of teaching form can be misleading. Many people, all over the world, "live" in their squat. They cook, clean, play, visit, and explore in a squat, which means that a squat is not a fixed thing. Once you've gotten your body used to squatting, try shifting your weight from side to side and front to back, all while down in your squat. There is not a right or wrong squat, although a systematic approach to form will assist you in making sure that you aren't doing squats that just reinforce the shape of your body that has already been troubling you in other ways.

Squatting is a critical component of a human's physical experience. Squats, done as exercise, can have a cumbersome feel—like taking an oversized multivitamin once a day that doesn't have much flavor. But once your body is ready, squats done in the context of all-day movement add other dimensions to your life. Communicating with nature (or little people) close-up, exercising the full birthright of your hip and knee motion, acquainting yourself with a digestive or procreation system at its best, are experiences I would hate for you, my fellow human, to be missing.

THE GREAT KEGEL DEBATE

While I have been a biomechanist since 2002, I wasn't a famous biomechanist until 2010, when I was interviewed for a blog called MamaSweat. In the interview, called "Pelvic Floor Party: Kegels are NOT invited," I gave ten answers to various questions on why women experience pelvic floor issues and what they should do about it. And no, my answers did not include "you should do Kegels," which is what the other tens of thousands of articles on the internet say you should do.

Not only did I not give Kegels as the solution, I also explained (in ten answers) the science behind why Kegels could very well be contraindicated for pelvic floor issues.

In the interview I mentioned that squatting ("as found in nature" squatting) was part of the pelvic floor's natural strengthening mechanism, as are a bunch of other things like muscle lengths and frequented pelvic positioning.

Overnight I became a sensation. In both a good and bad way. Over the course of a few months, this casual interview between friends was read eighty thousand times by men and women with pelvic floor issues and by the people who treat them. What I had written, while certainly heresy, was perceived and promoted (by some) as blasphemy.

The trouble with Kegel exercises is that they are an exercise prescription stemming from a particular understanding of muscle that doesn't consider, deeply, how muscle works. While I've written up the biomechanics of muscle force production in this book, specifically how chronic positioning and use cause muscle to shorten or lengthen at the level of the sarcomere, you won't find information like this in most anatomy textbooks. In most books and in most therapy research, muscle length is assumed to be the result of a rearrangement of sarcomeres, not in their increase or decrease. When your model is one of overlap, you could derive that a low-force-producing muscle must be one that is overstretched. Overstretched muscle needs to "tighten up" with contraction. However, it is also the case that a muscle with fewer sarcomeres has low force production (a difference that the tools used for measuring force in research cannot detect).

If you're using this oversimplified muscular model with regards to pelvic floor health, a Kegel is the natural solution. Only, as I was pointing out in the interview, it is just as likely that low force production could be the cause of muscles adapted (read: change in sarcomere number) to chronic pelvic and sacrum positioning, and in this case a Kegel would not only have little impact on the problem—it could exacerbate it as well.

I acknowledge that a ten-question interview was a poor medium through which to introduce the world to the notion that an exercise, prescribed to every woman via the pages of their favorite health magazine as both a preventative and corrective, could do harm. And had I known that eighty thousand people would be reading the article, I would have been more clear. And now that I have, let me clarify one more thing.

CONT'D ON NEXT PAGE

CONT'D FROM PREVIOUS PAGE

From the interview, many took my talk about the role squatting plays in pelvic floor health to mean that they should swap their Kegels for squats. As in, doing twenty or two hundred squats a day was the magic recipe for a perfectly functioning pelvic floor. As you now know from reading this book, pelvic floor issues—any issues, really—are the result of how you have moved your whole body over your entire life.

Your pelvic floor, like the TrA, has a wide range of "what it needs to do." Keeping it tight to "fix" one issue can disable its ability to perform another task naturally. Sure, you don't want a pelvic floor unable to generate force, but neither do you want one constantly generating too much tension. Your goal is a pelvic floor that performs its biological tasks—including the fun ones—superbly.

Like all muscles, performance requires the perfect amount of force for the task, which requires that a muscle be supple and used regularly. Meaning that you need to move your body, often, in lots of ways that call on the pelvic floor to respond in many unique ways. Let your pelvic floor play a full game of tennis, using a full range of shots. Don't set your pelvic muscles to SERVE and expect to win any games. (If that last reference doesn't make sense, go back and read the part on core activation and the TrA in chapter 8.)

So, from now on, when you ask me which exercises you need to do for a happy pelvis since Kegels "aren't good," the answer is: "Do all of them. All of the time. The end."

EPILOGUE

In elementary school I learned that caterpillars turn into butterflies, but it wasn't until I was watching a documentary on them as an adult that I learned *how* they did it.

In my child-mind, I had imagined that the butterfly parts were always packed into the container of the caterpillar, just waiting for the skin to molt off, exposing the butterfly that had been there all the time. But it turns out that caterpillars essentially dissolve themselves and grow all new parts. The process they undergo, the process of any being changing itself into an entirely different form, is called metamorphosis. The transition to a butterfly is profound, and the mechanism is so amazing, that the process of metamorphosis is often associated with spiritual transformation.

In schools all over the world, we hold up "animal kingdom" occurrences like these to illustrate for our children just how amazing animals are. A caterpillar gets wings! A snake sheds its entire skin! A sea star can grow additional arms! A chameleon changes its color! But while we're busy glorifying non-human animals, we're failing to recognize the human version of transformative processes. Our problem with nature is not only that we dwell outside of it, but that we see ourselves as separate from it, even when walking through it.

Our bodies changing in response to adaptations to our behavior may seem less drastic than suddenly sprouting wings, but if backing up your hips and lowering your rib cage throughout the day eliminates a chronic pain in your lower back, and walking differently, walking more, and sitting on the floor changes the distribution of mass in your bones and on your backside—aren't these changes just as profound?

Scientists are often criticized for their overinterest in detail, ostensibly because so many details can interfere with appreciation of the beauty found in nature. In one of my favorite interviews with Richard Feynman, a theoretical physicist, he discusses how an artist friend thinks Feynman, by breaking down the flower into how it works, reduces the beauty of it. Feynman disagrees, and states that, in addition to the beauty of the flower, "I see much more about the flower than he sees. I could imagine the cells in there, the complicated actions inside, which also have a beauty.…There's also beauty at smaller dimensions, the inner structure, also the processes.…It adds a question: does this aesthetic sense also exist in the lower forms? Why is it aesthetic? All kinds of interesting questions, [and] the science knowledge only adds to the excitement, the mystery, and the awe of a flower."

Humans don't seem to relate to their own beautiful mechanism in the same way as they relate to that of other creatures, probably due to a lack of awareness of happenings too small to see.

There is an Indian parable about perception. It goes something like this: A group of blind men stand around an elephant and are asked to identify what is in front of them. The man feeling the trunk describes the elephant as similar to a thick snake. Another at the tusk knows that the elephant is cold, hard, and sharp. A man at the side of the elephant describes a high, barrel-shaped object, one at the legs describes thick pillars, and another at the tail knows the elephant to be thin, flexible, and with a brush-like tip. In each case, the man speaks honestly— his truth created by his perception, but amongst each other, the men bicker and argue, leave insulting comments on each other's Facebook pages, and denounce each other for their ignorance.

The holistic perspective is one that considers every parcel to be true—each blind man speaks truthfully about what they "see"—but also acknowledges the mistake in assuming the parcel to be the whole truth. Stepping back to look at the entire elephant, and the entire environment in which the elephant dwells (and—Dr. Feynman—witness some of the elephant's biological processes at a microscopic level) changes what we see. It changes our version of the truth.

Making things even more complicated is that a single word, like *elephant*, can contain many parcels, many "true" definitions. If I say, "You're like an elephant," what am I saying, exactly? That you're tall? Gray? Have a huge nose? An excellent memory? Or that you take spectacular craps?

Ironically, it is words that most frustrate communication, especially when the involved parties are not aware of the mutable properties inherent in words. Like our bodies, words are shaped by our culture. Just as our preference for slouching has led to bucket-seat automobile and furniture designs that in turn result in our offspring slouching whether they prefer to or not (vicious cycle), words, shaped by culture, end up shaping culture right back.

As I've tried to demonstrate in this book, because exercise is the only experience we, as a sort of zoo animal, have had with movement, exercise becomes the only word we have to define "what humans do when they move their bodies about the planet." And thus, when we search for the answer to our problem, we've inadvertently limited the solutions available to us. When the word "movement" is eliminated from our vocabulary, so is the understanding of it.

And make no mistake, exercise and movement are not the only words that are currently shaping our relationship with wellness and the science of wellness. In his presentation "Biomedical Concepts and the Concept of Biological State," professor Stephen J. Lewis remarks:

> The reductionist approach that we have adopted has meant that the vast bulk of our knowledge about the body can, at best, be described as fractionated. Thus, words that have a holistic connotation—in that they refer to whole body states or experiences—have an inherently uncertain basis in science. The words illness, disease and health are like this.

When foundational words—words used every day, like illness, disease, and health—are without precise definitions, what, exactly, is everyone talking about?

Semantics, for some reason, is a science that seems to be often labeled as petty. And to include philosophical inquiries into words—well, some would say that one has veered entirely off the scientific track. But language is used to express and apply scientific findings; if we do not have a fixed definition of the words we use, the findings and the applications stemming from research are more prone to error.

When it comes to engineering, there is not an acceptable level of vagueness. If a construction company, contracted to build a bridge, interprets information differently from how it was intended by the architect, harm ensues. While not quite as catastrophic, our lack of specific definitions of terms used in human body research and, more importantly, terms used in resultant therapies, is slowly changing the quality of human study for the worse. Engineers use numerical statements to design safe buildings because numerical statements are extremely clear. We should strive for that same level of objectivity when we build sentences that eventually become the guidelines for human behavior.

As a science-geek-turned-writer, I have come to understand there really is never one "correct" definition for a word, as language evolves over time. I would, however, fail to honor the role of scientist if I didn't clarify key terms as I've used them here. This is why I've forgone a traditional glossary and put the definitions right in the text—so that you and I can be, literally, on the same page.

You have just read a book presenting a new paradigm that appears to be radically different from the paradigm we currently subscribe to. With further investigation you will find the difference between these two paradigms probably has less to do with scientific data and more to do with the vague language we have come to use regarding the body—even in academic settings. When you're looking at support material for existing points of view, always note the foundational definitions these documents hinge upon, rather than depending solely on the headlines.

Begin every thread of thought by asking yourself, "Which part of the elephant am I handling?" In conversations with colleagues, it is common to have discourse that begins with a "defend your position" attitude that turns into a discussion on definitions. "Oh, I wasn't using the word like that." Or, "The research *I'm* looking at didn't define the variable that way...or at all."

In no way do I believe that I have represented the broadest perspective in this book when it comes to human health, genetic expression, and physical outcomes. The complex interaction of thoughts, emotions, and beliefs on our structure (and language) is profound—and not even addressed on these pages. However, I *do* think that this book presents two or three huge steps away from the elephant in front of us and clarifies some of the issues keeping us from a more robust physical experience.

Because I've chosen to write a book for everyone, I've selected words that allow as many people as possible access to the information, but I might have selected different words if I were to present these ideas to you, personally. Every word here could have easily been expanded into ten, and every hunk of information presented in a chapter could easily be its own book. All this is to say that you might have questions now, because you didn't read *all* the information you needed.

At a recent presentation I attended, the speaker began her lecture with instructions to let information that didn't resonate with us fall to the ground like manure or compost, and to let information that did strike a chord germinate, like a seed.

As the seeds of what I have presented grow in you, begin further inquiry by going back before going forward. Explore old ideas with a new appreciation for language and definitions. Assess the foundation of your scientific understanding—the primary building blocks of the currently held tenets with respect to definitions—before integrating new information. Build the strongest possible edifice of knowledge that can expand in every direction.

While you are not a butterfly, by definition, you do transform over a lifetime, and it is just as amazing. It is my hope that I have highlighted some of the details behind your own beautiful, metamorphic process. It is also my hope that you see how you are not so different from the other animals and plants that coexist alongside of you. You are just as complex and adaptable, just as entwined with and affected by your environment, and just as capable of magical transformation.

Now, go ahead and move your DNA.

MOVE MORE

ALIGNMENT CHECKS

You're bound to find yourself standing around quite often—in line, waiting for the bus, or at a standing work station. While the end goal is to move your entire body more, making over the way you stand is an easy way to move your body differently for more advantageous loads. The following alignment checks are most fully applied to standing still, but also consider the pertinent checks when you're out walking or performing your correctives.

BEFORE

AFTER

KNEE PITS
NEUTRAL

ANKLES PELVIS-
WIDTH APART

FEET STRAIGHT

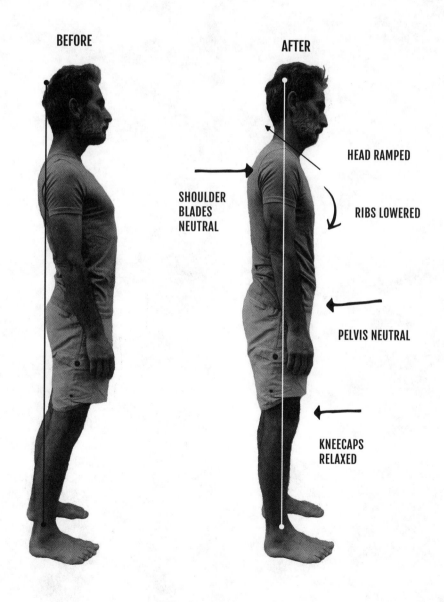

BEFORE

AFTER

HEAD RAMPED

SHOULDER
BLADES
NEUTRAL

RIBS LOWERED

PELVIS NEUTRAL

KNEECAPS
RELAXED

EXERCISE FLOWS

While my end goal is for you to be moving so much as a lifestyle the need for correctives disappears, they are key, at first, to get more of you moving. Here are three flows targeting different aspects of movement necessary to get your whole body moving more. Note, these flows don't include all the exercises; feel free to add or swap for any of your favorites, or create your own flows.

Restorative

Equipment Needs:
- Pillows (for sitting on)
- Half foam roller or rolled towel
- Yoga strap or belt
- Bolster or more pillows (as needed)

This is a flow of exercises designed to get your ankles, knees, hips, and shoulders moving in a gentle way. Always do exercises on both sides of the body (e.g., right leg and then left). Add bolsters (e.g., pillows under your hips or knees as necessary) to make these exercises more doable initially. Feel free to eliminate a move that's not working for your particular body. Unless otherwise noted, **perform each move for 30–60 seconds and most importantly, pay attention to the moves between moves—that is, how you transition from one position to another. This phase, too, is movement and can be improved.**

- Begin seated in a cross-legged position.

- Spend two minutes leaning forward and back and to the right and left, and finally twisting to the right and left

- Switch which leg is crossed on top and repeat

- Stand and do the Calf Stretch

- Top of the Foot Stretch

- Sit in Sole-to-Sole Sit. Spend two minutes leaning forward, twisting to the right and left

- Paint the Globe

- V-Sit, spend two minutes leaning forward, to the right and left, twisting to the right and left

- Stand and do the Calf Stretch

- Top of the Foot Stretch

- Floor Angels

- Spinal Twist

- Windmill

- Spinal Twist

- Floor Angels

- Calf Stretch

- Top of the Foot Stretch

- Prone Inner Thigh

- Quad Stretch

- Prone Inner Thigh

- Quad Stretch

- Cross-Legged Sit

- Reverse Prayer Hands

- Change legs in Cross-Legged Sit

- Finger Stretch

- Strap Stretch

- Psoas Release

You're done!

Foundational Strength

Equipment Needs:
- Pillows (for sitting on)
- Half foam roller or rolled towel
- Chair
- Yoga block or thick book

This is a flow of exercises designed to call on commonly underutilized muscles found in even the fittest of bodies! Always do exercises on both sides of the body (e.g., right leg and then left). Add bolsters (e.g., pillows under your hips or knees as necessary) to make these exercises more doable initially. Feel free to eliminate a move that's not working for your particular body. Unless otherwise noted, **perform each move for 30–60 seconds and most importantly, pay attention to the moves between moves—that is, how you transition from one position to another. This phase, too, is movement and can be improved.**

- Foot Bone Mobilization on Ball (spend 2–3 minutes per foot)

- Calf Stretch

- Top of the Foot Stretch

- Pelvic List

- Hand Stretch, Quadruped

- Rhomboid Pushup

- Lunge

- Double Calf Stretch

- Number 4 Stretch (seated or standing)

- Top of the Foot Stretch

- Pelvic List

- Rhomboid Pushup

- Paint the Globe

- Reverse Prayer Hands

- Lunge

- Patella Centering

- Reclined Sole-to-Sole Sit

- Patella Centering

- Reclined Cross-Legged Sit

- Patella Centering

- Rhomboid Pushup

- Double Calf Stretch

- Number 4 Stretch (seated or standing)

- Pelvic List

- Soles Against the Wall

- Strap Stretch

- Iliacus Release

- Foot Bone Mobilization on Ball

You're done!

Bigger Moves

Equipment Needs:
- Vertical/low bar
- Incline or rocks
- Monkey bars or thick tree branches

This is a flow of exercises designed to transition you to larger feats, like deep squatting, hanging, and swinging. I picture this routine done outside, to gain practice and the adaptations that come with ground-sitting and touching the ground with various parts of your body, as well as to allow you to use playground equipment or tree limbs for hanging and swinging. However, doing this flow indoors with the right equipment can help you transition to outdoor movement. You may find some exercises to be more doable than others. Add rest (in the form of a walking break), or modify or swap exercises as you find necessary, e.g., if you can't squat, swap a full squat for any squat prep exercise.

- Paint the Globe

- Vertical/low bar "hang," various grips

- Reverse Prayer Hands

- Supported Squat

- Calf Stretch au naturel

- Calf Stretch Soleus

- Supported Squat

- Paint the Globe

- Swinging, side to side

- Reverse Prayer Hands

- Supported Squat

- Cross-Legged Sit, moving front to back, side to side, twisting right and left

- Rhomboid Pushup

- On Your Back Squat

- Rhomboid Pushup

- Prone Squat

- Rhomboid Pushup

- Squat with Feet Bolstered

- Rhomboid Pushup

- Squat supported with arms

- Swinging front to back

- Calf Stretch

- Calf Stretch Soleus

- Cross the monkey bars as you can

You're done!

WALKING

Walking is a large category of movements, which means even if you already walk, there's a strong chance you could be moving more by changing how you're doing it.

Begin by quantifying your current walking habits using the questions below:

- If you use a pedometer or fitness tracking device, how many steps/miles do you average each day?
- Do you use a treadmill or walk overground?
- How many minutes do you walk each day?
- What rate do you typically walk (what's your average mile pace)?
- What route do you walk, and with what frequency?
- Note the terrain and grade that you frequent most.
- What shoes do you wear while walking?

There are two ways to increase the movement you do while walking: you can walk more and you can move more of you while walking.

Walk More

To walk more, you can either increase your amount of walking-for-exercise time or you can adjust your life a bit to walk more for the things you need (i.e., non-exercise walking). If it's the former, you can increase your walking gradually over time, keeping in mind that you may be able to add short walks, in the 2-15 minute range, every couple of hours more easily than you can add one or two hours of continuous walking to each day. To walk more for the things you need, consider these tips:

- Pick one or two places you go regularly that are less than a mile away, where you could walk, but often don't. This could be a coffee shop, the post office, or a friend's house. Resolve to always walk there (and if you're worried about time, this is a great time to catch up on phone calls, listen to your favorite music, visit with a family member, or just take some alone time).

- Drive *partway* to places you go regularly, and walk the rest. Maybe you're feeling places are too far to walk. To the office, to get the kids from school, or even to the grocery store—stop a bit short (whatever is feasible for you now) and walk the rest of the way. This is a sort of beefed up version of "park in the furthest parking spot." You know about that one already, right?

- Replace one typically sedentary social activity with a walking one. Meet a friend for coffee once a week? Get it to go and chat on the move. Replace your dinner and a movie date for a hike (#datehike). Swap out one reading session for an audiobook on the go. Instead of going to the craft supplies store, take your kids to the woods to gather free, biodegradable crafting materials (pebbles, sticks, pinecones, leaves, feathers, etc.).

Move More While You Walk

- **Do your correctives.** The corrective exercises in this book are designed to improve your mobility. Thus, once you've been doing them with regularity, you're likely to find changes to your gait (i.e., you'll be moving more while walking).
- **Choose different footwear.** Your shoes can be limiting how much of your body is able to participate while walking. Transition to shoes (minimal footwear) that allow your ankles and more joints in your feet to move. One note, though—more movement of these parts requires that you not only change your shoes, but where you're walking. So, with that in mind:
- **"Cross-terrain."** Start adding natural terrain frequently, even if it's just the bumpy grass and dirt next to the flat and cemented walking path at first. Small irregularities in the surfaces you're walking on add up to big changes—not only in your feet, but throughout your body.
- **Change your grade.** Add both big hills and smaller inclines to your walk to experience new loads and challenges, again, throughout the body. Take ramps instead of stairs or climb up the side of a hill vs. the less challenging meandering trail. Take stairs two at a time, just to mix things up.
- **Variability.** Benefit from the many advantages variability can give you by changing up the distance and frequency of your walks, your route, your speed, what you're carrying, who you're with, and where you go.

EXERCISE GLOSSARY

The following is a glossary of exercises with bullet points to help remind you of important form tips. You can use this section as a reference—to look up and practice individual exercises—or as a program itself. Running through each of these exercises daily, whether all at once or half in the morning and half in the evening, is one way to not only move more, but also to move more of you.

Unless indicated otherwise, aim to do each of these exercises two to three times, for thirty seconds to a minute every time. Start every exercise by assuming an aligned stance (see pages 226-227 for the alignment checks).

Abdominal Release

- Start on hands and knees, let your head hang, and allow your entire spine to relax, letting belly drop towards the floor.
- Once you feel you've released your diaphragm, try again—chances are you are holding residual tension there.

Calf Stretch

- Place the ball of your left foot on the top of a half foam roller or rolled-up folded towel or yoga mat, drop the heel all the way to the ground, and straighten that knee.
- Step forward with your right foot.
- If you can't bring your foot all the way forward, take a smaller step.
- Keep your weight stacked vertically over the heel of whichever foot is farther back.
- **Rainbow:** starting with the foot turned out 45°, step up onto a rolled towel (or dome) for a Calf Stretch. Turning in at 5–10° increments, load each joint angle via the basic Calf Stretch parameters.

Calf Stretch Soleus

- Starting from the position of the Calf Stretch, with the ball of one foot up on the dome and its heel on the ground, bend the same leg's knee—pushing it slightly forward—as you press that same heel toward the ground. Both knees can bend.

Cross-Legged Sit

- Sitting on the ground, cross one shin over the other, sitting up on a pillow or two as necessary.
- Try different combinations of leaning forward and rotating, as well as changing distance between shins and groin.
- Slowly "paint" an imaginary circle around you on the floor with your fingers.
- As you twist and lean forward, try to keep both sides of your bottom anchored to the floor.
- Repeat with other foot crossed in front.

Double Calf Stretch

- Stand facing the seat of a chair. With your feet pelvis-width apart, knees straight, and feet pointing forward, tip the pelvis forward until your palms rest on the chair.
- If you can't reach the chair without really bending the knees or rounding the back, add a pillow or stack of books to the seat until you can, or move to a counter or desktop.
- Once your arms are down, allow your spine to drop down toward the floor, and your tailbone to lift toward the ceiling.
- Don't force your ribs to the floor or arch your back—just relax the spine to the ground as much as you can.

Finger Extension

- Starting with hands palm up, extend wrists so that fingertips point toward the floor. Place fingertips on floor or tabletop, or use other hand to gently pull fingertips back.
- Gently press the heel of the hand forward, away from the body, as you bend your elbows until they point behind you (not to the side).
- If any finger joints are flexing, decrease the elbow bend.

Floor Angels

- Recline on a bolster or stacked pillows to allow your ribcage to release toward the floor.
- Reach your arms out to the side, thumbs toward the floor, and the elbows rotating up toward the ceiling.
- Lower your hands toward the floor, keeping your elbows slightly bent.
- Once your chest can handle this stretch, slowly move your arms toward your head, going as far as you can with your thumbs on the floor and while rotating the elbows toward the ceiling.
- Keep your ribs down (if they lift, you've gone too far).
- Make "snow angel" motions, moving slowly and going only as far as you can with your ribs down.

Foot Bone Mobilization on Ball

- Stand with a tennis or other squishy ball under the arch of one foot.
- Slowly load your weight (stay seated if necessary) onto the ball.
- Move your foot forward and back and side to side ("vacuuming" your foot) to gently articulate individual joints within the foot.
- **Rainbow**: Try different ball sizes and firmness, walk across cobblestone mats or make a cobblestone path to stand on daily.
- **Au Naturel**: As able, walk outside on natural terrain in bare or sock-covered feet.

Hand Stretch, Quadruped

- On your hands and knees (or against a wall works too), spread your fingers away from each other, trying for a 90-degree angle between middle finger and thumb.
- Straighten any bent finger joints and try to flatten any lifting palm-parts.
- Start rotating the upper arms so the elbow pits face forward, noticing (and correcting) any alterations to hand and finger placements.

Hanging Progression

Hanging Progression: Level-One Hanging

- Hold onto a vertical pole, low horizontal bar (or anything, really, that will support your body weight), keeping your feet near its base.
- Slowly extend your arm, dropping your upper body away from where you're holding. If your elbow joint is lax, maintain a slight bend.
- **Rainbow**: Try two hands, and change the distance of your feet, elevation of your hand(s), and hand position (palm up, down, etc.) for varying loads. Also try these first-level hangs on trees, where the bark can work to toughen up your hand skin!

Hanging Progression: Level-Two Hanging

- Find a bar or branch that allows you to keep your feet on the ground.
- Holding on with both arms, slowly bend your knees, dropping the weight of your body away from your hands to your feet.
- If your elbows straighten beyond 180°, put a small bend in them and work to maintain it. This will protect the ligaments in your elbows. If your shoulders end up around your ears, try to bring them down toward your waistband.
- Once you can stabilize your elbows and shoulders with a little help from your feet (this can take a month or two to develop), find a bar that allows you to hang with your feet off the ground.
- Progress to Swinging.

Iliacus Release

- Lie on your back, bolstering your ribs if necessary, with your knees bent.
- Prop the inferior (closer to your legs) half of the pelvis up on a bolster, yoga block, or stack of towels, making sure to leave space under your waistband.
- Like a teeter-totter, your pelvis should tip toward your head, lowering your waistband toward the floor.
- Don't work to rotate the pelvis; this defeats the point of "release." Allow gravity to create (or work on) this motion for you.

Legs on the Wall

- Sit sideways to the wall and rotate your body to get your legs straight up the wall, lying on your back.
- Move yourself away from the wall until your pelvis can rest with the ASIS and pubic symphysis aligned horizontally while your legs are up the wall.
- Keeping your legs straight, relax them away from each other until you feel a stretch in the inner thigh.
- Rest as necessary.

Lunge

- Start by sitting up on your knees (add padding as needed).
- Line up your pelvis—ASIS and pubic symphysis on a vertical plane.
- Step forward with your right foot and shift your weight forward toward the front foot as far as you can without your pelvis tilting.
- If you need to, scoot your front foot forward, until you feel your back leg can't extend any farther. Repeat other side.
- **Rainbow**: Begin by externally rotating the thigh bone of the back leg all the way (move your ankle toward the opposite side as far as it will go). Do a lunge here with your thigh in this position and again, each time repositioning your ankle (progressing away from the midline) three to four inches at a time.

Number 4 Stretch

- Sitting on the edge of a chair, bring your right leg up until your right ankle is resting on the left knee (sit on a pillow or two as necessary).
- Slowly tilt your pelvis and torso forward.
- For an additional challenge, do this stretch while standing. Initially using a desk or chair for balance, hook the ankle of one leg over the knee of the other, then lower your hips down while keeping the standing leg's knee directly above the ankle.

Paint the Globe

- Pretend your upper body is surrounded by a globe.
- Reach your arms up. That's the top of the globe.
- Reach down. There's the bottom.
- Reach your hands out to "touch" the sides of the globe.
- Now use your fingertips to "paint" as much of the globe's inner surface as you can.
- The more you do this, the easier it will become and the more of the globe you can touch.

Passive Prone Hip Extension

- Lie on your stomach and place a folded blanket or half foam roller under your pelvis.
- Relax your pubic symphysis to the floor while the ASIS stay supported on the pillow.
- Support upper body with forearms and keep ribs in line with pelvis.

Patella Centering

- Lie on your back with one knee bent and the other fully extended on the ground.
- Your pelvis should be positioned so the ASIS and pubic symphysis fall in a horizontal plane.
- Rotate the thigh of the straight leg until your knee pit is centered.
- Lift the straight leg slowly to the height of the opposite knee, without moving (watch for tucking!) the pelvis.
- Slowly lower leg to the floor with control.

Pelvic List

- Beginning with feet straight, ankles pelvis-width apart, weight on the rear of the foot, place your right hand on your right hip and pull the right side of the pelvis down toward the floor, bringing the left side up, hold for a moment, then lower it back down with control.
- Neither knee bends during this exercise.
- For more intensity, stand on a phone book or yoga block with the other leg floating above the ground. Repeat cues from above; the extra height increases the hip's range of motion, better preparing these muscles for walking hills.

Psoas Release

- Begin sitting on the floor, legs extended. Relax the muscles of the thighs until the hamstrings (muscles on the backs of the thigh) rest on the ground. You might have to lean back on your elbows to untuck your pelvis for this to happen.
- Once your thighs are down, recline, stopping just before the hamstrings lift away from the ground. At this angle, bolster your head and shoulders, leaving space for the ribs to lower to the floor.
- From here, relax the ribs toward the floor. As with all the other "releasing" exercises, the point is not to get your ribcage to the floor by contracting your muscles, but to recognize your habit of tensing muscles to lift it away and relax them.
- As you relax, your ribs will move closer to the floor. Continuous adjustment of the height or position of the bolster will be necessary.

Prone Inner Thigh

- Face down on the floor, resting your head on your hands, scoot one straight leg up toward your head, sliding it along the floor.
- Bring it up as high as you can with minimal lifting of one side of the pelvis away from the floor and without hiking one hip up toward your ribcage. Try to keep the right and left sides of your waist the same length.
- As you advance, try to externally rotate the stretching leg by rolling your thighbone so the toes on that foot turn toward the ceiling.

Quad Stretch

- Starting on your belly, place a bolster—a rolled-up yoga mat or towel works great for this—under the front of your pelvis (put your ASIS higher up on the roll) and let your pubic bone fall toward the floor.
- Bend your knee to bring your ankle to your hand without letting your pubic bone change position (i.e., don't let your pelvis tilt anteriorly).
- If you can't reach your foot, loop a strap (or a belt or tie) around your ankle to reach it with ease.
- Grab the lowest part of the shin and not the foot or toes. This will keep you from placing the load on hyper-lax ankle ligaments.

Reclined Sole-to-Sole Sit

- Lie on your back, bolstering the ribs as necessary. Bend your knees until they are close to your hips, then drop the knees out away from each other, keeping the soles of your feet touching.
- Place pillows under each knee to support you in this position as necessary. Lower the support bolsters as the position becomes more comfortable.
- **Rainbow**: Try this move varying up the distance between your feet and hips.

Reclined Cross-Legged

- From the position of the last exercise, cross your ankles to place one ankle in front of the other shin.
- Let both knees drop toward the floor, bolstering your knees as necessary.
- Repeat, crossing feet the opposite way.

Reverse Prayer Hands

- Place the backs of your hands together with all ten fingers touching (thumbs are the hardest).
- Lower your wrists until they're at the same height as elbows. Keep fingers together!

Rhomboid Pushup

- Begin on your hands and knees, with wrists below shoulders and knees below hips.
- Relax your head and spine.
- Elbow pits forward and arms straight (the elbows do not bend on this one), slowly lower your spine toward the floor, which will bring the shoulder blades together.
- Don't squeeze shoulder blades together; let gravity pull you down.
- After a moment at the "bottom," lift the entire spine up toward ceiling, spreading shoulder blades apart.
- Do not round upper back or tuck pelvis.

Shank Rotation

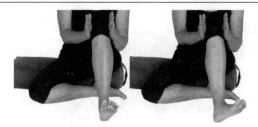

- Start seated with knees bent.
- Rotate the entire lower leg away from the midline of the body, which turns the foot outward.
- Try manually twisting your shin with your hands.
- Rotate inward (typically more difficult) and then work to turn outward.
- Keep the movement happening at the knee joint and not the ankle joint.

Sole-to-Sole Sit

- Sit on a pillow or folded blanket high enough to allow your ASIS to sit vertically above your pubic symphysis.
- Place soles of feet together, letting the knees drop away from each other.
- Tilt pelvis forward as if your pelvis is a bowl of soup you're pouring out in front of you.
- **Rainbow**: Lean forward to stretch your groin a bit more, as well as turn your torso to the right and left to change the load profile.

Soles Against a Wall

- Place the soles of both feet against a wall, making sure to elevate your hips until you can straighten your knees comfortably.
- Relax your body toward your thighs, no forcing, no bouncing.
- After relaxing your torso as far as it can go, relax your head toward your thighs.

Spinal Twist

- Lie on your back, both legs extended.
- Scoot your pelvis an inch or two to the right, bring the right knee up so that it stacks over your hip, and rotate your pelvis to lower that knee to the opposite side of your body, stopping as soon as your ribcage twists away from the ground.
- Twist only as far as you can without taking the ribcage with you—no forcing it.
- If you find that your pelvis barely moves and your knee is nowhere near the floor, stack pillows so that the knee crossing over can rest on them.
- Repeat on the other side.

SQUAT PREP EXERCISES

Squat Prep: On Your Back Squat

- Lie on your back and bring your knees in toward your chest.
- Check to see how much your pelvis tilted when you brought your knees in.
- Do it again, but this time keep your pelvis on the ground.

Squat Prep: Prone Squat

- Start on hands and knees, and ease back until your hips rest on your feet (or as far back as you can go).
- Come back to the start, and try the movement again, without tucking the pelvis.
- Add a pillow behind the knees or on top of the feet to help support your body weight during the stretch.
- Again, from your hands and knees, sit back again, only with your feet tucked to decrease the angle at the ankle. You can do this with your pelvis both tucked (more knee flexion) and untucked (less knee flexion).

Squat Prep: Squat with Feet Bolstered

- Place your heels up on a half foam roller, or a rolled yoga mat.
- Lower yourself into a deep squat.

Squat Prep: Squat Supported with Arms

- Hold on to something that can support your weight and step way back so that you're leaning forward.
- Lower yourself back into a squat, keeping shins vertical—holding on, but stepping back from support as necessary.
- Keep shifting your distance from the object until your arms are supporting you *just enough* to back your hips up and untuck your pelvis a bit more.

Strap Stretch

- Lie on your back with the hamstring of the right leg resting on the ground (if your hamstring doesn't touch, bolster your torso on pillows until it does).
- Wrap a strap (or belt) around the sole of the left foot, pulling the front of the foot toward you to stretch the calf.
- Lift the leg and fully extend the knee.
- Keeping the pelvis level, slowly bring the leg across the body (to the right in this case) until you feel the load in the lateral hip.
- Repeat other side.
- **Rainbow:** Rotate the thigh all the way out (look at your foot—it should be pointing all the way off to the side) before you bring the leg across the midline (image above right). Repeat this exercise, each time rotating your thigh in a bit (as noted by your foot position).

Swinging Progression

- Once you've mastered the hang (stable elbows and shoulders), see if you can get your body swinging forward and back and side to side.
- Try to work up to ten swings in each direction.
- **Rainbow:** You can use your legs, arms, and waist muscles to get your legs swinging. Try initiating the movements from each.

Top of the Foot Stretch

- Stand on your left foot and reach your right foot back behind you, tucking the toes of your right foot under and placing them on the floor.
- If you find yourself leaning forward, shorten the distance you've reached the leg back.
- Bring your pelvis over your standing ankle and upper body over the hips.
- Rest if you're cramping, and return to the move as you can.
- **Rainbow**: Drop the ankle side to side (holding at various points) and rotate the thigh to varying degrees before you press the ankle forward.

V-Sit

- Sitting up on a pillow or pillows as necessary, widen your legs as far as you comfortably can.
- Tilt your pelvis and torso forward, and hold.
- **Rainbow**: Lean side to side, twist both directions, and "paint" a rainbow shape with your hands on the floor.

Windmill Stretch

- Starting on your back, bring your left knee up toward your chest and then roll your entire body to the right (don't spinal twist, but roll) until your left knee rests on the ground.
- Align your ribcage to your pelvis (no rib thrust).
- Reach your left hand, arm, and shoulder blade up toward the ceiling and away from the spine.

- Slowly lower your arm to the left as far as you can without thrusting your ribcage (it's okay if it doesn't go to the floor) until you find the boundary of your tension.
- Once there, imagine your arm is on the face of a clock. Keeping your palm facing the ceiling, slowly move your arm down and up between twelve and six o'clock.
- Continuously reach your elbow away from your torso (not just your hand) and manage your ribs.
- Repeat on the other side.

APPENDIX: EXERCISE EQUIPMENT AND ADDITIONAL SOURCES OF INFORMATION

There are many places to obtain the equipment used in this book. Below are the places (or websites) from which I source my equipment for my home and studio. I've also included the recommended sizes and locations for easiest sourcing.

GENERAL EQUIPMENT

- **Bolster:** I recommend using a round bolster that is approximately 28 inches long with a 10-inch diameter. Find these sturdy cotton bolsters at your local yoga studio, or other online yoga supply companies like YogaAccessories.com.

- **Half foam roller**: I recommend using a 12x6x3 half foam roller (that's 12 inches long, 6 inches wide, and 3 inches tall). I order mine from foamerica.com. You can also find them on my website (nutritious movement.com/product/half-dome) as well as other exercise equipment stores.

- **Squat platform**: If you're interested in a squat platform for your toilet, you can find plans online to build one. If you'd like to purchase one already made, I recommend the Squatty Potty. You can often find these at Costco and Bed Bath & Beyond as well as online at squattypotty.com.

- **Yoga block**: If you're wanting to use a yoga block for the advanced pelvis list, I recommend one that's 9x6x3. You can likely find these at your local yoga studio, sports equipment store, Target, or online at yogaaccessories. com.

EQUIPMENT AND INFORMATION FOR YOUR FEET

Toe spacing devices

- **Foam toe separators**: Find inexpensive toe separators (typically used for pedicures) at any drugstore or pedicurist.

- **Correct Toes**: These toes spacers, which you can wear in shoes if your shoes are wide enough, are available at correcttoes.com.

- **Foot Alignment Socks**: These toe-spacing socks are available at my-happyfeet.com.

To move the bottom of your feet

- **Tennis ball:** Find at any sporting good store

- **Yoga Tune Up™ balls:** These are great for the Bottom of the Foot Stretch as they're soft and designed for just this activity. You can find them, as well as video instructions on how to use them, at yogatuneup.com.

ADDITIONAL INFORMATION SPECIFIC TO SHOES AND FEET

Find lists of minimal footwear on my website

- **Shoes: The List:** nutritiousmovement.com/shoes-the-list

My books on the matter

- *Simple Steps to Foot Pain Relief: The New Science of Healthy Feet*
  This is the more gentle introduction and focuses more on shoes.

- *Whole Body Barefoot: Transitioning Well to Minimal Footwear*
  This book has many more exercises as well as more technical information on anatomy and biomechanics.

MORE ON FASCIA

There are programs and products dedicated entirely to fascia and improving movement. I've listed some below; for more information, visit their websites:

- MELT Method: meltmethod.com

- Voodoo Floss Compression Bands: sold by roguefitness.com

- Yamuna Body Rolling: yamunausa.com

- Yoga Tune Up® Therapy Balls: yogatuneup.com

- *The Roll Model: A Step-by-Step Guide to Erase Pain, Improve Mobility, and Live Better in Your Body*, by Jill Miller, Victory Belt Publishing

The links below are for fascial therapies:

- Kinesis Myofascial Integration: anatomytrains.com/at/kmi
- Myofascial Release: myofascialrelease.com

MORE FOR YOUR PELVIC FLOOR

If you're currently experiencing pelvic floor issues, consider working with an alignment-aware physical or physiotherapist with additional training in pelvic floor issues. You can request information on their training in pelvic floor matters to see if they're the best fit for you.

Additional exercises, programming, and information I've created for pelvic alignment can be found in the following resources.

- In my exercise DVD *Nutritious Movement for a Healthy Pelvis* (find at nutritiousmovement.com).

- In my book *Diastasis Recti*, which, while dealing mainly with the core, also addresses pressures created through movement that constantly contribute to pelvic floor disorders.

OTHER WAYS TO LEARN FROM ME

Social media:

I offer regular, visual insights into a movement-rich life here: instagram.com/nutritiousmovement

Please note, my personal transition away from a sedentary life includes regular screen and social media breaks, and I've pruned my social media presence down to just Instagram. But there are years' worth of lessons to be found there, so peruse them as your own screen time allows to see which items resonate most with you.

My podcast:

You can listen to the award-winning *Move Your DNA* (another way to learn on the move) on iTunes, Spotify, Stitcher, or on my website nutritiousmovement.com, where you can also find transcripts of the show.

Exercise classes/videos:

Visit nutritiousmovement.com to find out more about the live and recorded classes I teach as well as my exercise DVDs and downloadable classes, or to find a movement teacher in your area.

RESOURCES AND FURTHER READING

INTRODUCTION

Al-Adawi, S. (2006). Emergence of Diseases of Affluence in Oman: Where do they Feature in the Health Research Agenda. *Sultan Qaboos University Medical Journal, 6*, 3-9.

Chen, C.S., and Ingber, D.E. (1999). Tensegrity and mechanoregulation: from skeleton to cytoskeleton. *Osteoarthritis Cartilage.*, 7(1), 81-94.

Ezzati, M., VanderHoorn, S., Lawes, C.M., Leach, R., James, W., Lopez, A., ... & Murray, C. (2005). Rethinking the "Diseases of Affluence" Paradigm: Global Patterns of Nutritional Risks in Relation to Economic Development. *PLOS Medicine 2*.

O'Keefe, J.H., Vogel, R., Lavie, C.J., & Cordain, L. (2010). Achieving hunter-gatherer fitness in the 21(st) century: Back to the future. *American Journal of Medicine, 123*(12), 1082-1086.

Wells, Calvin. (1964). *Bones, Bodies, and Disease; Evidence of Disease and Abnormality in Early Man*. Praeger: New York.

CHAPTER 1

Bloomfield, S.A. (1997). Changes in musculoskeletal structure and function with prolonged bed rest. *Medicine & Science in Sports Exercise, 29*, 197-206.

Cuccurullo, S., (Ed.) (2004). *Physical Medicine and Rehabilitation Board Review*. New York: Demos Medical Publishing.

Dettwyler, K.A. (1994). *Dancing Skeletons: Life and Death in West Africa*. Long Grove: Waveland Press, Inc.

Evans, W.A. (1996). Flaccid Fin Syndrome: Natural or Captive Phenomenon? (Masters thesis, Nova Southeastern University).

Feric, M., and Brangwynne, C. (2013). A nuclear F-actin scaffold stabilizes ribonucleoprotein droplets against gravity in large cells. *Nature Cell Biology, 15*, 1253-1259.

Hubmacher, Dirk, and Apte, Suneel S. (2013). The biology of the extracellular matrix: novel insights. *Current Opinion in Rheumatology, 25*(1), 65-70.

Nagle, K.B., and Brooks, M.A. (2011). A Systematic Review of Bone Health in Cyclists. *Sports Health 3*(3), 235-243.

Sage, E. Helene. (2001). Regulation of interactions between cells and extracellular matrix: a command performance on several stages. *Journal of Clinical Investigation, 107*(7), 781-783.

Scofield, K.L, and Hecht, S. (2012). Bone health in endurance athletes: runners, cyclists, and swimmers. *Current Sports Medicine Reports, 11*(6), 328-34.

Sherk, V.D., Barry, D.w., Villalon, K.L., Hansen, K.D., Wolfe, P., and Kohrt, W.M. (2014). Bone Loss Over 1 year of Training and Competition in Female Cyclists. *Clinical Journal of Sport Medicine 24*(4): 331-6.

Shull, P.B., Shultz, R., Slider, A., Dragoo, J.L., Besier, T.F., ... and Delp, S.L. (2013). Toe-in gait reduces the first peak knee adduction moment in patients with medial compartment knee osteoarthritis. *Journal of Biomechanics, 46*(1), 122-128.

Stewart, A.D., and Hannan, J. (2000). Total and regional bone density in male runners, cyclists, and controls. *Medicine and Science in Sports and Exercise, 32*(8), 1373-7.

CHAPTER 2

Abbott, A.L., and Bartlett, D.J. (2001). Infant motor development and equipment use in the home. *Child: Care, Health and Development, 27*(3), 295-306.

Active Healthy Kids Canada. (2013). Are We Driving Our Kids to Unhealthy Habits? The 2013 Active Healthy Kids Canada Report Card on Physical Activity for Children and Youth. Toronto: Active Healthy Kids Canada.

Aizawa, M., Mizuno, and Tamura, M. (2010). Neonatal sucking behavior: comparison of perioral movement during breast-feeding and bottle feeding. *Pediatrics International, 52*(1), 104-8.

American Society for Cell Biology. (2012, December 17). Breast cancer cells growing in 3-D matrix revert to normal. ScienceDaily.

Barry, H., and Paxson, L. (1971). Infancy and Early Childhood: Cross-Cultural Codes 2. *Ethnology, 10*(4), 466-508.

Batra, N.N., Li, Y.J., Yellowley, C.E., You, L., Malone, A.M., Kim, D.H., Jacobs, C.R. (2005). Effects of short-term recovery periods on fluid-induced signaling in osteoblastic cells. *Journal of Biomechanics, 38*(9), 1909-17.

Bril, B., Sabatier, C. (1986). The Cultural Context of Motor Development: Postural Manipulations in the Daily Life of Bambara Babies. *International Journal of Behavioral Development, 9*, 439.

Borghi, Nicolas, et al. (2012). E-cadherin is under constitutive actomyosin-generated tension that is increased at cell-cell contacts upon externally applied stretch. *Proceedings of the National Academy of Sciences 109.31*: 12568-12573.

Brownfield, D.G., Venugopalan, G., Lo, A., Mori, H., Tanner, K., Fletcher, D.A., Bissell, M.J. Patterned collagen fibers orient branching mammary epithelium through distinct signaling modules. *Current Biology 23*(8), 703-9.

Chaney, W.R. (2001) How Wind Affects Trees. *Indiana Woodland Steward, 10*(1).

Clarke, N.M.P. (2013). Swaddling and hip dysplasia: an orthopaedic perspective. *Archives of Disease in Childhood.*

Cole, W. G., Lingeman, J.M., and Adolph, Karen E. (2012) Go Naked: Diapers Affect Infant Walking. *Developmental Science, 15*(6), 783-790.

Eriksson, N., Benton, G., Chuong B., Kiefer, A., Mountain, J., Hinds, D., … & Tung, J. (2012). Genetic variants associated with breast size also influence breast cancer risk. *BMC Medical Genetics, 13.*

Hsieh, C. & Trichopoulos, D. (1991). Breast size, handedness and breast cancer risk. *European Journal of Cancer and Clinical Oncology, 27*(2), 131-135.

Ingber, D. (2003). Mechanobiology and diseases of mechanotransduction. *Annals of Medicine, 35*(8), 564-77.

Inoue, N., Reiko, S., and Kamegair, T. (1995). Reduction of masseter muscle activity in bottle-fed babies. *Early Human Development, 42*(3), 185-193.

Jacinto-Gonçalves, S.R., Gavião, M. B., Berzin, F., de Oliveira, A.S., Semeguini, T.A. (2004). Electromyographic activity of perioral muscle in breastfed and non-breastfed children. *The Journal of Clinical Pediatric Dentistry, 29*(1), 57-62.

Jorgens, D.M., Inman, J.L., Wojcik, M. et al. (2017). Deep nuclear invaginations are linked to cytoskeletal filaments—integrated bioimaging of epithelial cells in 3D culture. *Journal of Cell Science 130*(1): 177.

Kang, H.K., Kim, Y., Chung, Y., Hwang, S. (2012). Effects of treadmill training with optic flow on balance and gait in individuals following stroke: randomized controlled trials. *Clinical Rehabilitation, 26*(3), 246-55.

Lamb, Michael E., and Hewlett, Barry S., eds. *Hunter-Gatherer Childhoods: Evolutionary, Developmental, and Cultural Perspectives (Evolutionary Foundations of Human Behavior)*. Chicago: AldineTransaction, 2005.

Logan, D., Kiemel, T., Dominici, N., Cappellini, G., Ivanenko, Y., Lacquaniti, F., Jeka, J.J. (2010). The many roles of vision during walking. *Experimental Brain Research, 206*(3), 337-50.

Martin, R. Bruce, Burr, David B., & Sharkey, Neil A. *Skeletal Tissue Mechanics.* New York: Springer-Verlag, 1998.

Milstein, J.N., and Meiners, J.C. (2011). On the role of DNA biomechanics in the regulation of gene expression. *Journal of the Royal Society Interface, 8*(65), 1673-1681.

Mulavara, A.P., Richards, J.T., Ruttley, T., Marshburn, A., Nomura, Y., Bloomberg, J.J. (2005). Exposure to a rotating virtual environment during tread-mill locomotion causes adaptation in heading direction. *Experimental Brain Research, 166*(2), 210.

Orr, W. A., Helmke, B.P., Blackman, B.R., Schwartz, M.A. (2006). Mechanisms of Mechanotransduction. *Developmental Cell, 10*(3), 407.

Prokop, T., Schubert, M., Berger, W. (1997). Visual influence on human locomotion. Modulation to changes in optic flow. *Experimental Brain Research, 114*(1), 63-70.

Raichlen, D.A., Pontzer, H., Harris, J.A. et al. (2016). Physical activity patterns and biomarkers of cardiovascular disease risk in hunter-gatherers. *American Journal of Human Biology* Oct 9. doi: 10.1002/ajhb.22919. [Epub ahead of print]

Schneider, Robert, and Grosschedl, Rudolf. (2007). Dynamics and interplay of nuclear architecture, genome organization, and gene expression. *Genes & development 21.23,* 3027-3043.

Surovell, T.A. (2000). Early Paleoindian Women, Children, Mobility, and Fertility. *American Antiquity, 65*(3), 493-508.

Shirai, N., Shigaru, I. (2012). Reduction in sensitivity to radial optic-flow congruent with ego-motion. *Vision Research, 62,* 201-208.

Siegel, A.C., and Burton, R.V. (1999). Effects of baby walkers on motor and mental development in human infants. *Journal of Developmental Behavioral Pediatrics, 20*(5), 355-61.

Stock, J.T. (2006). Hunter-gatherer postcranial robusticity relative to patterns of mobility, climatic adaptation, and selection for tissue economy. *American Journal of Physical Anthropology, 131,* 194-204.

Toyota, M. & Gilroy, S. (2012). Gravitropism and mechanical signaling in plants. *American Journal of Botany, 100*(1), 111-125.

Wall-Scheffler, C.M., Geiger, K. and Steudel-Numbers, K.L. (2007). Infant carrying: The role of increased locomotory costs in early tool development. *American Journal of Physical Anthropology, 133,* 841-846.

Yoshiko, Y., Gentaro, T. (2010). Influence of experience of treadmill exercise on visual perception while on a treadmill. *Japanese Psychological Research, 52*(2), 67-77.

CHAPTER 3

Brew, B.K., Marks, G.B., Almqvist, C., Cistulli, P.A., Webb., K. Marshall, N.S. (2014). Breastfeeding and snoring: a birth cohort study. *PLoS One, 9*(1).

Glazebrook, K.E. and Brawley, L.R. (2011.) Thinking about maintaining exercise therapy: Does being positive or negative make a difference? *Journal of Health Psychology 16*(6): 905–916.

Gourgou, S., Dediu, F., & Sancho-Garnier, H. (2002). Lower Limb Venous Insufficiency and Tobacco Smoking: A Case-Control Study. *American Journal of Epidemiology, 155*(11), 1007-1015.

Kannus, R., Józsa, L, Renström, R., Järvotoen, M., Kvist, M., Lento, M., ... & Vurol, I. (1992). The effects of training, immobilization and remobilization on musculoskeletal tissue. *Scandinavian Journal of Medicine & Science in Sports, 2*(3), 100-118.

Kawano, H., Tanimoto, M., Yamamoto, K., Sanada, K., Gando, Y. Tabata, I., ... & Miyachi, M. (2008). Resistance training in men is associated with increased arterial stiffness and blood pressure but does not adversely affect endothelial function as measured by arterial reactivity to the cold pressor test. *Experimental Physiology, 93*(2), 266-302.

Khan, K.H., Scott, A. (2009). Mechanotherapy: how physical therapists' prescription of exercise promotes tissue repair. *British Journal of Sports Medicine, 43,* 247-252.

Kohli, M.V., Patil, G.B., Kulkarni, N.B., Bagalkot, K., Purohit, Z., Dave, N., Sagari, S.G., Malaghan, M. (2014). *Journal of Clinical and Diagnostic Research, 8*(3), 199-201.

Limme, M. (2010). The need of efficient chewing function in young children as prevention of dental malposition and malocclusion. *Archives de pédiatrie, 17*(5), S213-9.

Lopes, T.S., Moura, L.F., Lima, M.D. (2014). Association between breast-feeding and breathing pattern in children: a sectional study. *Jornal de pediatra,* doi: 10.1016/j.jped.2013.12.011

Noreen von Cramon-Taubadel. (2011). Global human mandibular variation reflects differences in agricultural and hunter-gatherer subsistence strategies. *PNAS 108*(49), 19546-19551.

CHAPTER 4

Alberts, B., Johnson, A., Lewis, J., et al. (2002). Genesis, Modulation, and Regeneration of Skeletal Muscle. In *Molecular Biology of the Cell,(4th ed)*. New York: Garland Science.

Bertovic, D.A., Waddell T.K., Gatzka, C.D., Cameron, J.D., Dart, A.M., and Kingwell, B.A. (1999). Muscular strength training is associated with low arterial compliance and high pulse pressure. *Hypertension, 33*(6), 1385-91.

Boonyarom, O., and Inui, K. (2006). Atrophy and hypertrophy of skeletal muscles: structural and functional aspects. *Acta Physiologica (Oxford)*, *188*(2), 77-89.

Clifford, P.S. (2007). Skeletal muscle vasodilatation at the onset of exercise. *The Journal of Physiology*, *583*(Pt 3), 825-833.

Conway, D., and Schwartz, M. (2013). Flow dependent cellular mechanotransduction in atherosclerosis. *Journal of Cell Science*, *126*, 5101-9.

Hademenos, G.J., and Massoud, T.F. (1998). *The Physics of Cerebrovascular Diseases: Biophysical Mechanisms of Development, Diagnosis and Therapy*. College Park, MD: American Institute of Physics.

Haga, J.H., LI, Y.S., and Chien, S. (2007). Molecular basis of the effects of mechanical stretch on vascular smooth muscle cells. *Journal of Biomechanics*, *40*(5), 947-60.

Humphrey, J.D., ed. (2002). *Cardiovascular Solid Mechanics: Cells, Tissues, and Organs*. New York: Springer Verlag.

Humphrey, J.D. (2008). Vascular adaptation and mechanical homeostasis at tissue, cellular, and sub-cellular levels. *Cell Biochemistry and Biophysics*, *50*(2), 53-78.

Khan Academy. (2012). *Sarcomere Length-Tension Relationship*. [video and slides] Retrieved from: http://youtu.be/uVFqEi5j1v0.

Mann, C.J., Periguero, E., Kharraz, Y., Aguilar, S., Pessina, P., Serrano, A.L., and Muñoz-Cánoves, P. (2001). Aberrant repair and fibrosis development in skeletal muscle. *Skeletal Muscle*, 1:21.

Morton, J.P., Kayani, A.C., McArdle, A., Drust, B. (2009). The exercise-induced stress response of skeletal muscle, with specific emphasis on humans. *Sports Medicine*, *39*(8), 643-62.

Tardieu, C., Tabary, J.C., Tabary, C., and Tardieu, G. (1982). Adaptation of connective tissue length in immobilization in the lengthened and shortened positions in cat soleus muscle. *The Journal of Physiology, 78,* 214-217.

Thayer, S. (2010). *Nature's Garden: A Guide to Identifying, Harvesting, and Preparing Edible Wild Plants.* Ogema: Forager's Harvest Press.

Vial, C., Zúñiga, L.M., Cabello-Verrugio, C. Cañón, P., Fadic, R., Brandan, E. (2008). Skeletal muscle cells express the profibrotic cytokine connective tissue growth factor (CTGF/CCN2), which induces their dedifferentiation. *Journal of Cellular Physiology, 215*(2), 410-21.

Yamamoto, K., Kawano, H., Gando, Y., Iemitsu, M., Murakam, H., Sanada, K., … & Miyachi, M. (2009). Poor trunk flexibility is associated with arterial stiffening. *Heart and Circulatory Physiology, American Journal of Physiology, 297*(4), 1314-8.

CHAPTER 5

Birn-Jeffery, A.V., and Higham, T.E. (2014). The Scaling of Uphill and Downhill Locomotion in Legged Animals. *Integrative and Comparative Biology.* [Epub ahead of print]

Morgan, Christopher. (2008). Reconstructing prehistoric hunter-gatherer foraging radii: a case study from California's southern Sierra Nevada. *Journal of Archaeological Science, 35*(2), 247-258.

Seireg, A. and Arkivar, R.J. (1975.) The prediction of muscular load sharing and joint forces in the lower extremities during walking. *Journal of Biomechanics* 18: 89–102.

Venkataraman, V.V., Kraft, T.S., Desilva, J.M., and Dominy, N.J. (2013). Phenotypic plasticity of climbing-related traits in the ankle joint of great apes and rainforest hunter-gatherers. *Human Biology, 85*(1-3), 309-28.

Venkataraman, V.V., Kraft, T.S., and Dominy, N.J. (2013). Tree climbing and human evolution. *Proceedings of the National Academy of Sciences of the United States of America.*, 110(4), 1237-42.

CHAPTER 6

Barreto de Brito, L., Ricardo, D., Soares de Araújo, D., Ramos, P., Myers, J., and Soares de Araújo, C. (2012). Ability to sit and rise from the floor as a predictor of all-cause mortality. *European Journal of Preventive Cardiology*, 21(7), 892-898.

Bergmann, G., et al. 1995. Influence of shoes and heel strike on the loading of the hip joint. *Journal of Biomechanics, 28* (7), 817-27.

Chevalier, G., Sinatra, S., Oschman, J., Sokal, K., and Sokal, P. (2012). Earthing: Health Implications of Reconnecting the Human Body to the Earth's Surface Electrons. *Journal of Environmental and Public Health, 2012.*

D'Août, K., et al. 2009. The effects of habitual footwear use: Foot shape and function in native barefoot walkers. *Footwear Science, 1*(2), 81-94.

Hewes, G.W. (1955). World Distribution of Postural Habits. *American Anthropologist* (April), 231-244.

Nurse, M.A., et al. (2005). Changing the texture of footwear can alter gait patterns. *Journal of Electromyography and Kinesiology, 15*(5), 496-506.

Raichlen, D.A., Wood, B.M., Gordon, A.D., Mabulla, A.Z., Marlowe, F.W., and Pontzer, H. (2014). Evidence of Levy walk foraging patterns in human hunter-gatherers. *Proceedings of the National Academy of Sciences, 111(2),* 728-33.

Rome, K., Hancock, D., and Poratt, D. 2008. Barefoot running and walking: The pros and cons based on current evidence. *The New Zealand Medical Journal, 121*(1272).

Rossi, W.A. (1999). Why shoes make "normal" gait impossible. *Podiatry Management* (Mar.), 50-61.

Rossi, W.A. (2001). Footwear: The primary cause of foot disorders. *Podiatry Management* (Feb.), 129-38.

Warden, S.J., Burr, D.B., and Brukner, P.D. (2006). Stress fractures: Pathophysiology, epidemiology, and risk factors. *Current Osteoporosis Reports, 4*(3), 103-109.

CHAPTER 7

Crockett, H., Gross, B., Wilk, K., Schwartz, M., Reed, J., O'Mara, J., ... & Andrews, J. (2000). Osseous Adaptation and Range of Motion at the Glenohumeral Joint in Professional Baseball. *The American Journal of Sports Medicine, 30(1)*, 20-26.

Kraft, T.S., Venkataraman, V.V., Dominy, N.J. (2014). A natural history of human tree climbing. *Journal of Human Evolution*, 71, 105-18.

Peterson, J. (1998). The Natuflian hunting conundrum: spears, atlatls, or bows? *International Journal of Osteoarchaeology, 8*(5), 378-389.

CHAPTER 8

Alton, F., et al. 1998. A kinematic comparison of overground and treadmill walking. *Clinical Biomechanics, 13*(6), 434-40.

Bogduk, N., Pearcy, M., & Hadfield, G. 1992. Anatomy and biomechanics of psoas major. *Clinical Biomechanics, 7*(2), 109-19.

Cordain, L., Eaton, S., Brand Miller, J., Lindeberg, S., & Jensen, C. (2002). An evolutionary analysis of the aetiology and pathogenesis of juvenile-onset myopia. *Acta Ophthalmologica Scandinavica, 80*(2), 123-35.

Dancy, M., Christian, W., & Belloni, M. (2002). *The Human Eye*. Retrieved from http://webphysics.davidson.edu/physlet_resources/dav_optics/examples/eye_demo.html

Diab, M. (1999). *Lexicon of Orthopaedic Etymology*. Amsterdam: Harwood Academic Publishers. 276-277.

Horowitz, S.S. (2013). *The Universal Sense: How Hearing Shapes the Mind*. New York: Bloomsbury USA.

Hu, H., et al. (2011). Is the psoas a hip flexor in the active straight leg raise? *European Spine Journal, 20,* 759-65.

Jones-Jordan, L., Sinnott, L., Cotter, S., Kleinstein, R., Manny, R., Mutti, D., ... & Zadnik, K. (2012). *Investigative Ophthalmology & Visual Science, 53,* 7169-7175. http://dx.doi.org/10.1167/iovs.11-8336

Kirchmair, L., et al. 2008. Lumbar plexus and psoas major muscle: Not always as expected. *Regional Anesthesia and Pain Medicine, 33* (2), 109-14.

Krause, B. (2001 Rev. 2006). *Loss of Natural Soundscape: Global Implications of Its Effect on Humans and Other Creatures*. [transcribed speech]. Retrieved from http://www.escoitar.org/2006-08-23-Loss-of-Natural-Soundscape-Global-Implications-of

Miller, J. (1974). Effects of noise on people. *The Journal of the Acoustical Society of America, 56,* 729.

Morgan, K. & Tromborg, C. (2007). Sources of stress in captivity. *Applied Animal Behavior Science, 102*(3-4), 262-302.

Mutti, D. (2013). Time outdoors and myopia: a case for Vitamin D? *Optometry Times*. Retrieved from http://optometrytimes.modernmedicine.com/optometry-times/news/time-outdoors-and-myopia-case-vitamin-d

National Digestive Diseases Information Clearinghouse (NDDIC), U.S. Department of Health and Human Services. (2013). *Digestive Diseases Statistics for the United States.* Retrieved from http://digestive.niddk.nih.gov/statistics/statistics.aspx

National Institutes of Health, U.S. Department of Health and Human Services. (2009.) Opportunities and Challenges in Digestive Diseases Research: Recommendations of the National Commission on Digestive Diseases. Bethesda, MD: National Institutes of Health. NIH Publication 08–6514.

Saw, S., Chua, W., Wu, H., Yap, E., Chia, K. & Stone, R. (2000). Myopia: gene-environment interaction. *Annals Academy of Medicine Singapore, 29(3)*, 290-7.

Tetley, M. (2000). Instinctive sleeping and resting postures: An anthropological and zoological approach to treatment of low back and joint pain. *British Medical Journal, 321*(7276), 1616-18.

CHAPTER 9

Alton, F., Baldey, L., Caplan, S., & Morrissey, M.C. (1998). A kinematic comparison of overground and treadmill walking. *Clinical Biomechanics, 13*(6), 434-440.

Barnett, C.H. (1954). Squatting facets on the European talus. *Journal of Anatomy, 88*(Pt 4), 509-513.

Branko. (2006). *Clap Skate.* [drawing]. Retrieved from http://en.wikipedia.org/wiki/File:Clap_skate.svg

Callaghan, J.P., Aftab, E.P., McGill, S.M. (1999). Low back three-dimensional joint forces, kinematics, and kinetics during walking. *Clinical Biomechanics, 14*(3), 203-216.

Carpinella, I., Crenna P., Rabuffetti, M., and Ferrarin, M. (2010). Coordination between upper- and lower-limb movements is different during overground and treadmill walking. *European Journal of Applied Physiology, 108*(1), 71-82.

Coenen, P., van Werven, G., van Nunen, M.P., Van Dieën, J.H., Gerrits, K.H., and Janssen, T.W. (2012). Robot-assisted walking vs overground walking in stroke patients: an evaluation of muscle activity. *Journal of Rehabilitation Medicine, 44*(4), 331-7.

Hausdorff, J.M. (2007). Gait dynamics, fractals and falls: finding meaning in the stride-to-stride fluctuations of human walking. *Human Movement Science, 26*(4), 555-89.

Kong, P.W., Koh, T.M.C., Tan, W.C.R., and Wang, Y.S. (2012). Unmatched perception of speed when running overground and on a treadmill. *Gait & Posture, 36*(1), 46-48.

Kuo, A.D. (2007). The six determinants of gait and the inverted pendulum analogy: A dynamic walking perspective. *Human Movement Science, 26*(4), 617-56.

Lindsay, T.R., Noakes, T.D., and McGregor, S.J. (2014). Effect of treadmill versus overground running on the structure of variability of stride timing. *Perceptual and Motor Skills, 118*(2), 331-346.

Mohler, B.J., Thompson, W.B., Creem-Regehr, S.H., Pick Jr., H.L, and Warren Jr., W.H. (2007). Visual flow influences gait transition speed and preferred walking speed. *Experimental Brain Research, 181*(2), 221-228.

Multon, F., and Olivier, A-H. (2013). Biomechanics of Walking in Real World: Naturalness we Wish to Reach in Virtual Reality. In Steinicke, F., Visell, Y., Campos, J., Lécuyer, A. eds., *Human Walking in Virtual Environments*, 55-77. New York: Springer.

Parvataneni, K., Ploeg, L., Olney, S.J., Brouwer, B. (2009). Kinematic, kinetic and metabolic parameters of treadmill versus overground walking in healthy older adults. *Clinical Biomechanics, 24*(1), 95-100.

Prosser, L.A., Stanley, C.J., Norman, T.L, Park, H.S., and Damiano, D.L. (2011). Comparison of elliptical training, stationary cycling, treadmill walking and overground walking. Electromyographic patterns. *Gait & Posture, 33*(2), 244-50.

Seiler, S. (1997). The New Dutch "Slapskates": Will They Revolutionize Speed Skating Technique? *Sportscience News.* Retrieved from http://www.sportsci.org/news/9703/slapskat.htm

Steele, L. (2014). Fit People Don't Just Run, They Move. *Outside.* Retrieved from http://www.outsideonline.com/fitness/bodywork/in-stride/Fit-People-Dont-Just-ExerciseThey-Move.html

Thomson, A. (1889). The Influence of Posture on the form of the Articular Surfaces of the Tibia and Astragalus in the different Races of Man and the Higher Apes. *Journal of Anatomy and Physiology, 23*(Pt 4), 616-639.

Tskhovrebova, L., and Trinick, J. (2010). Roles of titin in the structure and elasticity of the sarcomere. *Journal of Biomedicine and Biotechnology,* doi: 10.1155/2010/612482. [Epub].

Zeh, E.P., and Duysens, J. (2004). Regulation of arm and leg movement during human locomotion. *Neuroscientist, 10*(4), 347-61.

CHAPTER 10

Cho, Y.K., Kim, C.S., Koo, E.S., Yun, J.W., Kim, J.W., Lee, J.H., ... & Choi, D.I. (2003). Contribution of Posture to Anorectal Angle and Perineal Descent on Defecography. *Korean Journal of Gastroenterology, 41*(3), 190-5.

Choi, J.S., Wexner, S.D., Nam, Y.S., Constantinos, M., Sallum, M., Yamaguchi, T., ... & Yu, C.F. (2000). Intraobserver and interobserver measurements of the anorectal angle and perineal descent in defecography. *Diseases of the Colon & Rectum, 43*(8), 1121-1126.

Gupta, N.P., Kumar, A., & Kumar, R. (2008). Does Position Affect Uroflowmetry Parameters in Women? *Urologia Internationalis, 80,* 37-40.

Kidd, R.S., & Oxnard, C.E. (2002). Patterns of morphological discrimination. *American Journal of Physical Anthropology, 117(2),* 169-181.

Lambiase, R.E., Levine, S.M., Terek, R.M. & Wyman, J.J. (1998). Long bone surface osteomas: Imaging features that may help avoid unnecessary biopsies. *American Journal of Roentgenology, 171(3),* 775-8.

Petros, P.P. & Skilling, P.M. (2001). Pelvic floor rehabilitation in the female according to the integral theory of female urinary incontinence: First report. *European Journal of Obstetrics & Gynecology and Reproductive Biology, 94(2),* 264-269.

Poli de Araujo, M., Takano, C.C., Girão, M.J., Gracio, M. & Sartori, F. (2009). Pelvic floor disorders among indigenous women living in Xingu Indian Park, Brazil. *International Urogynecology Journal, 20(9),* 1079-1084.

Sakakibara, R., Tsunoyama, K., Hosoi, H., Takahashi, O., Sugiyama, M., Kishi, M., ... & Yamanishi, T. (2010). Influence of Body Position on Defecation in Humans. *Lower Urinary Tract Symptoms, 2,* 16-21.

Sikirov, B.A. (1990). Cardio-vascular events at defecation: Are they unavoidable? *Medical Hypotheses, 32*(3), 231-3.

Sikirov, D. (2003). Comparison of Straining During Defecation in Three Positions: Results and Implications for Human Health. *Digestive Diseases and Sciences, 48*(7), 1201-1205.

Singh, A. (2007). Do we really need to shift to pedestal type of latrines in India? *Indian Journal of Community Medicine, 32*, 243-4.

Wang, K. & Palmer, M. (2010). Women's toileting behaviour related to elimination: Concept analysis. *Journal of Advanced Nursing, 66*(8), 1874-1884.

EPILOGUE

Lewis, Stephen J. *Biomedical Concepts and the Concept of Biological State* [transcript and PowerPoint slides]. Retrieved online from https://sites.google.com/site/sjlewis55/presentations/temah2

INDEX

ABOUT THE AUTHOR

photo by: Jen Jurgensen

Bestselling author, speaker, and a leader of the Movement movement, biomechanist Katy Bowman, M.S. is changing the way we move and think about our need for movement. Her ten books, including the groundbreaking *Move Your DNA*, have been translated into more than sixteen languages worldwide.

Bowman teaches movement globally and speaks about sedentarism and movement ecology to academic and scientific audiences such as the Ancestral Health Summit and the Institute for Human and Machine Cognition. Her work is regularly featured in diverse media such as the Today Show, CBC Radio One, the *Seattle Times*, NPR, the Joe Rogan Experience, and *Good Housekeeping*.

One of Maria Shriver's "Architects of Change" and an America Walks "Woman of the Walking Movement," Bowman consults on educational and living space design to encourage movement-rich habitats. She has worked with companies like Patagonia, Nike, and Google as well as a wide range of non-profits and other communities to create greater access to her "move more, move more body parts, move more for what you need" message.

Her movement education company, Nutritious Movement, is based in Washington State, where she lives with her family.